Bodyspace

Anthropometry, Ergonomics and Design

Bodyspace

Anthropometry, Ergonomics and Design

Stephen Pheasant

Taylor & Francis
London · New York · Philadelphia
1988

UK Taylor & Francis Ltd, 4 John Street, London WC1N 2ET

USA Taylor & Francis Inc., 1900 Frost Road, Suite 101, Bristol, PA 19007

Reprinted 1988, 1990

British Library Cataloguing in Publication Data

Pheasant, Stephen
Bodyspace: anthropometry, ergonomics and design.
1. Anthropometry
I. Title
620.8′2 GN51
ISBN 0-85066-340-7 (hbk)
ISBN 0-85066-352-0 (pbk)

Library of Congress Cataloging-in-Publication Data is available on request

Cover design by Keith Morton

Typeset by Alresford Typesetting and Design, Alresford, Hants

Printed in Great Britain by Burgess Science Press, Basingstoke, Hants

Foreword

We are so used to seeing people of widely varying dimensions that, unless a person of extreme height or girth appears, we take it all for granted. So, too, do we accept the dimensions of the various things we use; doorways are always high enough (for most of us), chairs and tables we accept and use, accepting too the not infrequent discomfort arising from their use. The assumption—unrecognized perhaps—behind this acceptance is that it cannot be otherwise, it is up to us to adapt to suit the things we use, and that it is not really very important anyway.

The increasing amounts of time that more and more people spend sitting, both at the office and in motor vehicles, and the widespread incidence of low back pain in the population, have caused many to question this simple assumption regarding the unimportance of relating things more closely to people. Early studies in the field of ergonomics looked at the reaches needed for control panels and workplaces of many kinds, and seating was an early subject of study. Anthropometry, previously a study developed for the classification and identification of racial differences and the effects of diet, living conditions, etc., on growth, was pressed into service to provide information about human dimensions around which workplaces could be designed. Initially many of the decisions were simple: the length of a straightened arm from shoulder to clenched fist was the reach distance, the length of the forearm gave information about areas of easiest reach, whilst the distance from the underside of the thigh to the ground, when the leg at the knee was bent to a right angle, was the appropriate dimension for the height of a chair seat.

These and similar dimensions, collected from many different populations over the last four decades, still provide many people with their material for decisions about workplace designs. Yet, as this book demonstrates, the data and their use have become more complex. Ergonomists recognize more clearly the importance of quite a close match to people's needs. Small variations in the height of work activities in relation to the seat, even of one or two centimetres, can induce or eliminate pains in the neck or shoulders. A small forward inclination of the trunk, maintained over a period, can be more uncomfortable, and perhaps as damaging, as more obviously extreme postures. The jobs that people do are now recognized as being as important in the specification of the dimensions of a workplace as how tall they are.

This point, that the purposes which people are trying to achieve are major influences on the way anthropometric data are applied, is the relatively new view which makes this book so timely. When we knew little about the effects of work on behaviour, we were satisfied to make simple decisions. With the increased ability to measure more reliably more of the results of these decisions, we understood more about the subtle interactions between people and their workplaces—and, of course, the other areas where they spend their lives. The results of this increased understanding can be pressed into service to improve the lot of vast numbers of people. The improvement will not be hailed by the national media as a major scientific step for a man—or for woman either for that matter—indeed, it may pass entirely unnoticed. The effects, however, will be profound for the increasing numbers of the world's population involved in the use of furniture, equipment, tools, vehicles and all those many objects which are part and parcel of an industrialized society. These effects will include such crucial matters for all of us as reduced musculoskeletal disease, less discomfort and better performance at work and a more active and comfortable old age.

We may confidently expect continuing developments in anthropometry and in its applications, and this book is an important milestone in that direction. The problems of the use of anthropometry as a tool subject have here had the full consideration that they now need to enable designers, ergonomists and many others to exploit the results. It is by the many engaged in making the world's artefacts more suitable for its population that this book will be welcomed.

E. N. Corlett

Contents

Acknowledgements

I should like to express my thanks to Professor Maurice Bonney who supplied Figure 2.13, and Communication Complex Design (CCD) Ltd. for permission to use the mannikin designs in Figures 2.9–2.12. I should particularly like to thank Keith Morton who drew all the remaining figures which show any signs of artistic talent (those which show no such talent are my own responsibility). The Department of Education and Science gave permission to publish the data in Tables 4.8–4.23 and 4.39. I should also like to thank Rosemary Townend who patiently typed and re-typed the manuscript and the staff of Taylor & Francis Ltd who brought the project through to its conclusion.

This book is dedicated to all my friends and loved-ones who have patiently given me their support during the time of its preparation under circumstances which were more difficult than we ever anticipated.

He who would do good to another must do it in Minute Particulars:
General Good is the plea of the scoundrel, hypocrite and flatterer,
For Art & Science cannot exist but in minutely organised Particulars,
And not in generalising Demonstrations of the Rational Power.

William Blake, *Jerusalem*, pl. 55, 1.60–64.

Chapter 1
Introduction

> Several similar contests with the petty tyrants and marauders of the country followed, in all of which Theseus was victorious. One of these evil doers was called Procrustes or the stretcher. He had an iron bedstead on which he used to tie all travellers who fell into his hands. If they were shorter than the bed he stretched their limbs to make them fit; if they were longer than the bed he lopped off a portion. Theseus served him as he had served others.
>
> From *The Age of Fable* by Thomas Bullfinch (1796–1867)

To state that we inhabit a predominantly artificial world is perhaps a truism. None the less, it is all too easy to ignore the simple fact that most of the visible and tangible characteristics of the environments in which we spend the greater parts of our lives, are the consequences of design decisions. It is unlikely that you are reading this in a desert wilderness; more probably you are indoors surrounded by furniture, or in a moving vehicle, or at least in a cultivated garden. The decisions which lead to the creation of these artificial environments may or may not be made by design professionals. They may be the consequences of extensive planning or of momentary whims—but they represent choices which have been made, which could presumably have been made otherwise, but were not by any means inevitable. All too often the artefacts which we encounter in our man-made environment are like so many Procrustean beds to which we must adapt.

The thesis of this book is that if an object or a space is intended for human use, then its form and dimensions should be derived from those of the human body, from the characteristics of the human senses and from the verifiable data of human experience. In other words: this book is about *bodyspace*.

1.1. *Some definitions*

If you were to look up the word 'ergonomics' in a dictionary, you might find it defined as "the scientific study of human beings in relation to their working environments". This traditional definition will serve us well enough, provided that its elements are treated in a sufficiently broad sense. The concept of work must encompass a wide range of human behaviour—not only in the tasks performed in the occupational context but leisure and domestic activities as well. Similarly, our study of the working environment must include not only the physical environment and the objects within it (such as machines, furniture, tools, etc.) but also psychological factors such as mental workload, the flow of information and social interactions with other human beings. A consideration of the multiple exchanges between human and environment may lead us to think in terms of man–machine systems or, on a broader scale, of socio-technical systems. By way of illustration, a person (of either sex) operating a computer terminal would be an example of a man–machine system, whereas many such people and machines functioning within an organisational context would be a socio-technical system. The 'systems concept' has been very influential in the development of ergonomics, as has the idea of the man–machine interface (or more generally the user interface). This may be represented as an imaginary surface across which information is transmitted from machine to operator by means of displays and from operator to machine by means of controls—in the case of our computer operators the screen and keyboard, respectively (Figure 1.1).

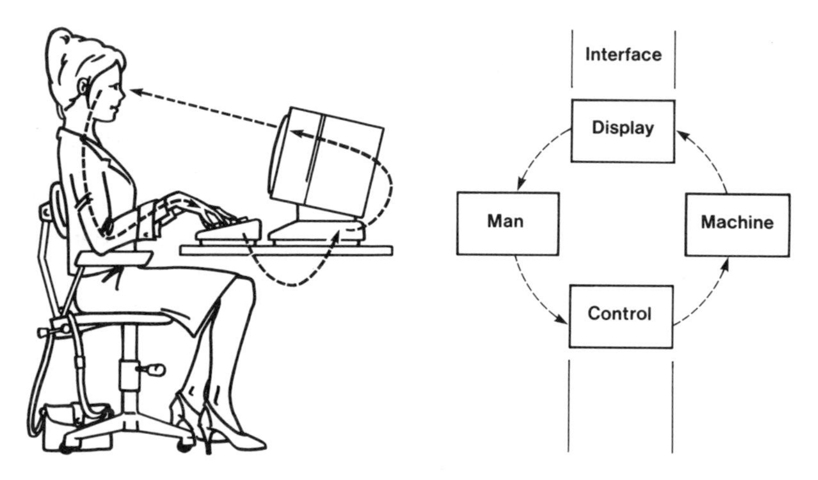

Figure 1.1. The 'man–machine interface'. Information passes from user to machine via controls and from machine to user via displays.

The most obvious reason why we should undertake this study of the relationships between human beings and the artefacts and environments which they use (other than simple curiosity) is with the intention of changing things for the better—either so that the performance, productivity, health or safety of the user may be improved or simply to make the user's experience a more pleasing and satisfying one. We might attempt this for altruistic reasons or, more probably, because it makes sound commercial sense to do so. In the occupational context we may choose to call this 'fitting the job to the man' and perhaps to compare it with 'fitting the man to the job' (by selection, training, vocational guidance, etc.). The two approaches are commonly seen as mutually interdependent.

Ergonomists today mostly tend to think of themselves as technologists or engineers rather than as scientists. To put the distinction plainly, a scientist studies things or analyses them, an engineer makes things or causes them to be made and a technologist is something of both. Indeed, a discipline indistinguishable from ergonomics evolved in the USA under the name 'Human Factors Engineering', 'Human Engineering' or 'Human Factors'. Historically, both disciplines originated as 'technological fall-out' from World War II and were applied firstly in the military context, secondly in the industrial context and subsequently expanded into more general areas of application. Although a few people would see a distinction between 'Ergonomics' and 'Human Factors' the two terms are now being used interchangeably on either side of the Atlantic.

Inherent in the process of 'changing things for the better' is the concept of design—whether this be the design of a physical object, a working method, an environment or a system. So for our present purposes let us propose the following definition:

> Ergonomics is the application of scientific information about human beings (and scientific methods of acquiring such information) to the problems of design.

The present book makes no claim to cover the science and technology of ergonomics in its entirety. Our principal concern will be with the particular area of the subject known as 'applied anthropometry' or 'anthropometrics'; that is with numerical data concerning the sizes, shapes and other physical characteristics of human beings and with the application of these data in the design context. (But we shall branch out from this central theme when the occasion demands.)

Anthropometry in general may be simply defined as the measuring of human beings. Its historical antecedents may be traced back, through the works of Renaissance artists and authors, to classical times. The *Four Books of Human Proportions* of Albrecht Durer (1471–1528) are the beginning of scientific anthropometry. Durer attempted to characterize the diversity of human physical types, and his exquisite illustrations are, according to his own assertion, based on the systematic observation and measurement of substantial numbers of people. Durer stands at the watershed between 'modern' empiricism and an earlier 'classical' tradition which sought to propose a 'canon of human proportions' describing the well-made human body according to certain pre-established aesthetic principles. By an argument based on an analogy with the observable properties of musical instruments, simple whole-number relationships between the dimensions of the human body were considered 'harmonious'. The Roman architectural theorist Vitruvius, writing sometime around the year 15 B.C., argued that since these proportions represented beauty in a very fundamental sense, they could be applied in the design of buildings. Two thousand years ago anthropometry and design were considered to be related—but the reasoning underlying the relationship is unappealing to the twentieth-century mind. In the Renaissance this theory of aesthetics became very influential. The famous drawing by Leonardo da Vinci (1452–1519), in which a man's body is shown inscribed within a square and a circle, is derived directly from Vitruvius. It is on the classical prescriptive side of the watershed, rather than the modern descriptive one (it is also so well-known an image that it scarcely requires repetition here). Curiously enough, the classical tradition has re-emerged in the present century in the work of the French architect Le Corbusier (1887–1965). His definitive statement, *The Modulor—A Harmonious Measure to the Human Scale Universally Applicable to Architecture and Mechanics* (Le Corbusier 1961), is an obscure work considered by many to be profound. Theories of human proportion have been treated from an art historical standpoint by Panofsky (1970)—their broader significance in the history of ideas would merit further discussion.

The development of an empirical science of anthropometry was closely linked, particularly in the nineteenth and early twentieth centuries, with physical anthropology and was particularly concerned with attempts to subdivide and classify the human race according to the physical characteristics of its members. More recently, human growth and the classification of physiques have been important foci of interest; medicine and sport being important areas of application. Although we shall touch on these subjects in Chapter 3, they are not central to our theme and will only be discussed to the extent that they are relevant to ergonomics and design.

What then is the relationship between ergonomics, anthropometry and design? The definitions which were proposed above place anthropometry as a basic human science contributing to ergonomics, which in turn contributes data, concepts and methodologies to the design process (Figure 1.2). Under ideal conditions the flow

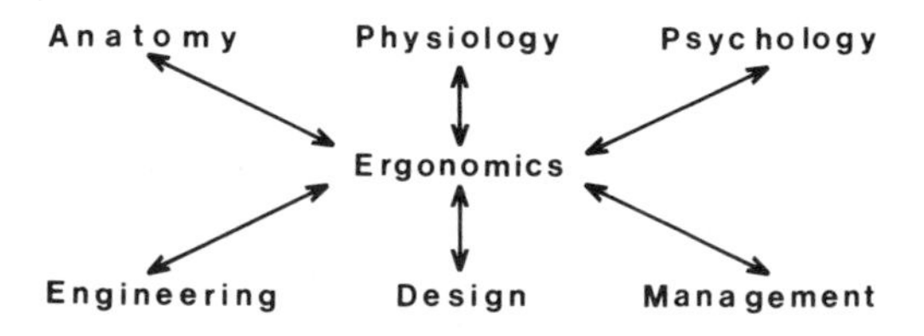

Figure 1.2. Ergonomics as an information channel.

of information should be in both directions—since practical applications in design and engineering should pose new problems for the basic sciences to investigate.

Anthropometric factors are by no means the only ones with which ergonomics (and the ergonomist) deals; none the less, they are frequently of fundamental importance in the match between operator and machine or user and product. Let us again consider the computer-terminal operators that we mentioned at the outset. Their basic bodily comfort at the terminal and whether or not they suffer backache, stiff necks, headaches, etc., will be determined in no small measure by anthropometric considerations (see Chapter 8)—together with factors such as lighting (environmental ergonomics). Their effectiveness within the man–machine system will be determined also by factors such as the linguistic design of computer dialogues and the ordering of system functions (software or cognitive ergonomics). The operational effectiveness of the man–machine system will depend on its organizational context (systems ergonomics)—and so on through the various hierarchical levels of system complexity. (Most ergonomists are not inclined to venture further up this hierarchy than the design of relatively small socio-technical systems—few of us would admit to political aspirations.)

The effective solution of problems in each successive hierarchical level presupposes at least a relatively successful match in the levels below—a computer operator with a blinding headache is unlikely to contribute much to the overall smooth operation of the system, however 'user-friendly' its software.

Since the solution of every design problem is in some sense a unique creative act we may reasonably suppose that its human factors will be similarly unique. Hence, there is no single correct solution to the problem of 'an ergonomically designed office chair'; rather there are a whole range of possibilities, each more or less satisfactory under certain circumstances. Ergonomics is not only an assembly of data; more importantly it provides a way of looking at the process of design itself. Its viewpoint may be characterized as 'from the human user outward', or 'user-centred', and this provides us with our final definition:

> Ergonomics is the scientific foundation, both in terms of data and methodology, for a user-centred approach to design.

1.2. *User-centred-design—a scientist in the glasshouse*

A well-designed object should be structurally sound, functionally appropriate (e.g., convenient and safe to use) and aesthetically pleasing. Few people would argue with these criteria which can, in fact, be found in Vitruvius, who named the desirable qualities in buildings as Utilitas, Firmitas and Venustas. The first two terms are self-explanatory, the last is approximately rendered as 'beauty'. In most social contexts the object should also be designed in such a way that it can be manufactured and sold at a profit. Clearly, these desirable features are by no means independent—rather they overlap. A tool which is constructed of inappropriate materials (e.g., a screwdriver with a soft metal blade) becomes unsafe and inconvenient to use precisely because it is structurally unsound—and it will not sell in a market of informed consumers. Similarly, the comfort of a chair is both a functional and an aesthetic characteristic—if we allow the term 'aesthetic' to apply to all of the senses rather than restricting its use to vision alone. (Neuroanatomists tell us that vision accounts for around 40% of the incoming sensory input to the brain—or in other words 40% of our sensory experience. The ways in which a chair may give tactile and kinaesthetic pleasure are, it seems to me, comparable with its appearance if we step back a little from the designer's traditional preoccupation with visual experience.)

To some extent, of course, all design is user-centred, and any designer (or manufacturer) who totally ignores the needs of the users of his products will rapidly become unemployable (or bankrupt). It is often easy to be critical and difficult to be constructive—and sometimes it seems that this condition is exacerbated by a scientific education, the emphasis of which must necessarily be the critical appraisal of experimental evidence and the hypotheses based thereupon. In the passages that follow, I shall run a considerable risk of conforming to the stereotype of the scientist, descending from his ivory tower and standing in a glasshouse vigorously throwing stones in all directions.

None the less, it does seem that there is a sense in which much contemporary design and architecture fails to concern itself with the real needs of the user—when one hears discussion of the sculptural properties of a hairdryer what should one conclude?

This divorce from the end user is particularly curious in the context of a theory of design called 'functionalism'. The proposal that creative decisions concerning the form of an object should be determined by its function has been central to the development of twentieth-century architecture and design—particularly via the Bauhaus school and the 'International Modern' style of architecture. But if we take such modern classics as the Marcel Breuer 'Wassily' chair (1925) or the Mies van der Rohe 'Barcelona' chair (1929) we find very little relationship between the form of these seats and that of the human body which it is (presumably) their function to support. The fact that such pieces are commonly referred to as 'occasional chairs' implies that they are without particular function—except to be used 'occasionally'. (In fairness we must admit that the Barcelona chair was in fact designed for the King of Spain to sit on at the opening of an exhibition.)

If we look back to earlier periods of furniture design, for example to the early years of the eighteenth century in Britain, we find a very different state of affairs. The William and Mary, Queen Anne and early Georgian periods produced furniture in general, and chairs in particular, which showed a closeness of functional relationship with the human body which has never been excelled (Figure 1.3). Consider the William and Mary winged chair and the variety of ways in which it may provide the postural support necessary for relaxation; or the Queen Anne dining chair (sometimes known as the Hogarth chair) with its gently curved back which reflects the form of the human spine. Neither should we ignore those furniture types of the Georgian period designed for various very specific functions indeed—the library or 'cock-fighting' chairs which gentlemen would sit astraddle, the feminine equivalent for kneeling upon, the reading stands and even the 'night table' on which to empty the contents of the pockets. All these bespeak a paramount concern for user requirements—a relationship between maker and user which is also apparent in much vernacular design (perhaps most clearly so in the hand tools used by woodworkers and other craftsmen).

At some time around the midpoint of the eighteenth century, we see function gradually playing an increasingly accessory role as design was dominated by a succession of aesthetic theories or 'styles'—neo-classicism, Gothic, etc. Para-

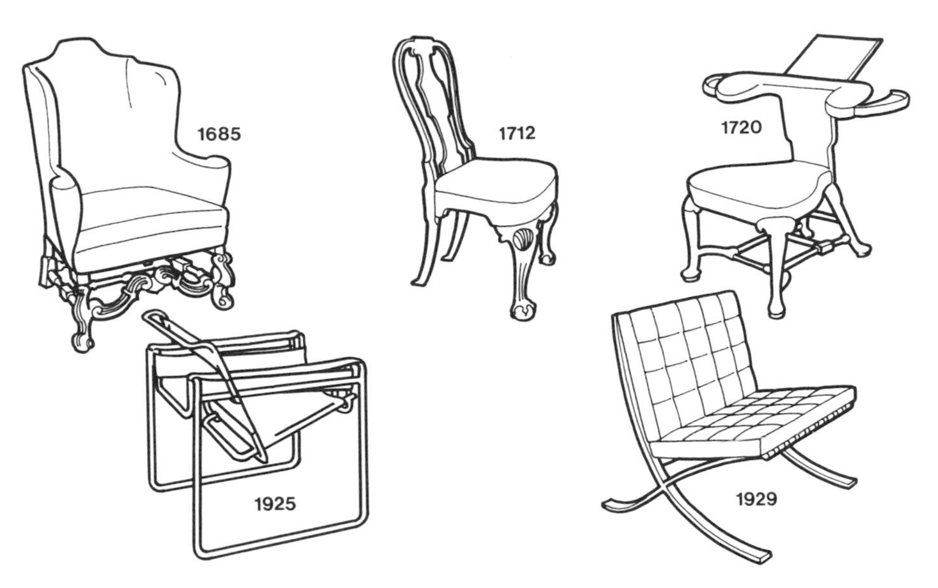

Figure 1.3. Form and function—eighteenth century style and twentieth centurey style. (Upper row: left to right) William and Mary winged armchair, Queen Anne dining chair, early Georgian library chair. (Lower row) 'Wassily' chair by Marcel Breuer, 'Barcelona' chair by Mies van der Rohe. For a contrast see also Figure 11.3.

doxically, the most recent of these styles is called 'functionalism', but it should be seen as an aesthetic demand for absence of ornament, 'truth to materials', etc., rather than a particular concern with end use. Functionalism is essentially a visual metaphor by which a designed object may acquire certain desirable connotations.

We may detect in product design and architecture a preoccupation with the visual experience of the user at the expense of other concerns. If we turn our attention to industrial engineering we may very well find little consideration of the user at all. A classic study from the early industrial years of ergonomics satirized the design of a capstan lathe by suggesting that its ideal operator should be 4½ ft tall with an armspan of 8 ft. Keeping the working parts of the lathe as close to the ground as possible makes excellent mechanical sense, but does little for the operator's spine (see Chapter 8). Pictures of this individual are to be found, for example, in Singleton (1963). He has come to be known as 'the Cranfield man', since that was where he was first described. By way of an alternative, Figure 1.4 shows the ideal operator of certain computer terminals. It is my sincere hope that she will not come to be known as 'the Hampstead woman'.

The stones which have been scattered in our ergonomically designed glasshouse are by now getting quite numerous, but there are a few more left to throw as we conclude our consideration of user-centred design and its alternatives by discussing 'the five fundamental fallacies' (Table 1.1). The reader is asked to believe that I have heard all of these statements from intelligent professional people,

Figure 1.4. Ideal operator for certain desk-top computers. She is provided with an extra eye at the level of her xiphisternum—"all the better to view the screen with".

although they are not, of course, always expressed in so many words. They are concerned with two themes. The first is the contrast between the subjective or intuitive approach to problem solving, which typifies the creative work of the designer, and the deductive methods, based on empirical data, of the scientist. Lawson (1980) described consistent differences between students of architecture and science in the approaches they adopted in a problem-solving task. The architects' strategy was described as 'solution-focused' and the scientists' as 'problem-focused'. The architects "learnt about the nature of the problem largely as a result of trying out solutions, whereas the scientists set out specifically to study

Table 1.1. The five fundamental fallacies.

No. 1	This design is satisfactory for me—it will, therefore, be satisfactory for everybody else.
No. 2	This design is satisfactory for the average person—it will, therefore, be satisfactory for everybody else.
No. 3	The variability of human beings is so great that it cannot possibly be catered for in any design—but since people are wonderfully adaptable it doesn't matter anyway.
No. 4	Ergonomics is expensive and since products are actually purchased on appearance and styling, ergonomic considerations may conveniently be ignored.
No. 5	Ergonomics is an excellent idea. I always design things with ergonomics in mind—but I do it intuitively and rely on my common sense so I don't need tables of data.

the problem". Now there is nothing whatever wrong with a 'solution-focused' approach—provided that its end results are properly evaluated (preferably before they are passed on to the user). The second theme is the question of 'Human Diversity'. In the author's view this is the single most important characteristic of people to be borne in mind in the world of practical affairs in general and of design in particular. To put it plainly, people come in a variety of shapes and sizes—to say nothing of their variability in strength, dexterity, mentality and taste. As we shall see the five fallacies are increasingly difficult to refute.

Not many people would express the first fallacy in so many words, but in implicit form it is very widespread. How many products are actually tested at the design stage on a representative sample of users? More commonly the evaluation of a design proposal is entirely subjective. The designer considers the matter, tries out the prototype and concludes that it 'feels alright to me', with the clear implication that if it is satisfactory 'for me' it will be for other people too. In general, objects designed by the stronger or more able members of the population can create unsurmountable difficulties for the weaker and less able. Women frequently say with exasperation "You can tell it was designed by a man!"

The first fallacy is closely linked with the last by the concept of empathy, of which more anon; it is also closely linked to the second since most people consider themselves to be more or less average. Suppose we were to determine the dimensions of a door by the average height and breadth of the people who were to pass through it. The 50% of people taller than average would bang their heads; the 50% wider than average would have to turn sideways to squeeze themselves through. Since the taller half of the population are not necessarily the wider half, we would, in fact, satisfy or accommodate less than half of our users. Nobody would make such an elementary mistake in designing a door—but, in my experience, the second fallacy turns up quite frequently in the work of students, of both design and ergonomics, who have only partially grasped the principles of anthropometrics. Obviously enough, we must seek to accommodate the largest percentage possible of the user population (see Chapter 2).

The third fallacy really has the ring of truth. Human beings are indeed very adaptable—they will put up with a great deal and might not necessarily complain. In the example we have just quoted, the taller half of the population would presumably learn to duck. This is the Procrustean approach to design. Adaptation to the Procrustean bed commonly has 'hidden costs' in terms of ill health, although only rarely are these as dramatic as an amputated limb. Consider the economic losses occasioned by the extensive range of musculoskeletal disorders which may be attributed to faulty workspace design.

Part of the refutation of the third fallacy rests upon these hidden costs of adaptation. In addition we should consider that the design process not only responds to consumer needs but in some measure creates them as well. We could question the extent to which (*a*) the public gets what the public wants, (*b*) the public wants what the public gets or (*c*) the public knows perfectly well what it wants, but can't get it and puts up with whatever is available. Superimposed over these possibilities are the effects of marketing and advertising on the one hand and consumer pressure groups and legislation on the other. The designer must respond to a variety of socio-economic forces and his artefacts reflect the society in which they are created. In some cases consumer pressure leads to the introduction of ergonomic features into design—this is happening quite dramatically at the present time in the area of office technology. The terminals of today are ergonomically better than those of a decade ago, probably because of the effects that user pressure (particularly through trade unions) has had upon the balance of market forces. In some marketing situations consumers are prepared to pay extra for quality. In a later section we shall consider the desirability of providing kitchen worksurfaces at a range of heights—this is perfectly possible technically but is generally deemed uneconomic. For which 'quality' would the informed consumer rather pay extra—an elegant finish with gleaming worktops and polished brass door furniture or ease of use and less backache? However, beyond all these considerations is the simple fact that making something the right size is often no more expensive than making it the wrong size. The decision to ignore ergonomics on grounds of economics is often just an excuse.

The fifth and final fallacy involves some rather complex issues. The intuition and common sense which we speak of in this context is sometimes called 'empathy'—and if you are a designer you may well have it in abundance. (Whether it is an innate gift or the fruit of experience is another matter.) Empathy is an act of introspection or imagination by which we may 'place ourself in another person's shoes'. It could be argued that, by empathetically casting oneself in the role of the user, the act of designing for others becomes an extension of designing for oneself and the traditional subjective approach becomes valid. In some measure this is probably true, but can these intuitions really circumvent the problems of human diversity? Can we really imagine how somebody quite different from ourselves would experience a certain situation? As far as I am aware, the hypothesis has not been directly tested, although as a psychological question it has important ramifications. It must surely be very difficult for a fit young adult male, for example, to place himself in the shoes of an elderly arthritic trying to get out of an armchair or the harassed mother of three dragging her boisterous offspring around the supermarket. In such cases empirical data must surely be more reliable however intuitive we consider ourselves to be. Common sense is in itself a difficult concept to analyse. At one level expressions like "that's just common sense" can be used as a justification for the blind acceptance of an untested hypothesis. Cameron (1984) has characterized common sense as good practice in problem solving. It can be argued that common sense and the scientific method are one and the same—the latter representing a refined and organized version of the former. To gather as much information about the user population as possible and to test one's intuitions objectively is good practice in problem solving. The scientist has cast all of his stones—he must now gather them together for some constructive purpose—or else return to his ivory tower.

Chapter 2
Principles and practice of anthropometrics

In the previous chapter attention was drawn to the desirability of allowing for human diversity in the design of equipment and environments. We must now turn our attention to how this may be achieved both in principle and in practice. There are situations in which workspaces and accoutrements may be designed specifically for an individual user; bespoke tailoring, haute couture and the customized seats used by racing drivers are cases in point. These could be characterized as luxury goods. Most of us are prepared to accept 'off the shelf' solutions, which approximately match our physical characteristics and would be disinclined to pay the extra for a perfect fit, even if we supposed it could be achieved in practice. For a few of us, the supposed 'luxury' of custom design becomes a necessity if we are to have any independent existence at all—the physical characteristics and requirements of the severely disabled are so variable that aids to mobility or independence must often be specially made for the individual.

We all acknowledge the necessity of manufacturing garments in a range of sizes, but would it also be true to say that chairs and tables, for example, should be supplied in a range of sizes as well? The answer is "only to a limited extent". We do not expect adults and children to use the same sized writing desks in their offices and schools; although they seem to cope perfectly well with the same dining table at home. We commonly supply typists with adjustable chairs; but their desks are usually of fixed height. Obviously enough, we are prepared to accept a less accurate fit from a table and chair than from a shirt and trousers. What is rather less obvious is how we should choose the best compromise dimensions for equipment to be employed by a range of users and at what point we should conclude that adjustability is essential. In order to optimize such decisions we require three types of information:

(*i*) the anthropometric characteristics of the user population;

(*ii*) the ways in which these characteristics might impose constraints upon the design;

(*iii*) the criteria which define an effective match between the product and the user.

Before discussing this matter further it is necessary to establish the quantitative foundations upon which our subject rests. I have attempted to do this with the least possible recourse to mathematical equations. This should suffice for the general reader; the specialist may refer to Section 2.7 for a more rigorous treatment.

2.1. *The statistical description of human variability*

In order to establish the statistical concepts which describe human variability, let us conduct what earlier scientific writers would have called an experiment of the imagination. Supposing you are in a large public building frequented by a fairly typical cross-section of the population. A companion, who is an inveterate gambler, offers to take bets on the stature (standing height) of the next adult man to walk down the corridor. (We could just as well bet on women, children or everyone taken together—but it is a little easier to deal with the problem mathematically, if we only consider adults of one sex.) On what height would you be best advised to place your money (assuming of course that you have no prior knowledge of people who happen to be in the area)? You will probably pick a stature which is somewhere near the average, since experience has told you that middling sized people are relatively common whilst tall or short people are rare by comparison. You have in essence made a judgement as to the relative probability of people of different statures, or the relative frequency with which such people are encountered by chance. Average people are more probable than extremes, in that you encounter them more frequently. The statistically minded punter, offered a bet of this kind, could optimize his chances of winning by going out and measuring all the men in the building. With these data we could plot a chart like the one shown in Figure 2.1, in which probability (frequency of encounter) is plotted vertically against stature, horizontally. The smooth curve on this chart is known as a probability density function or a frequency distribution. The particular curve we have drawn here is symmetrical about its highest point—the average stature, otherwise

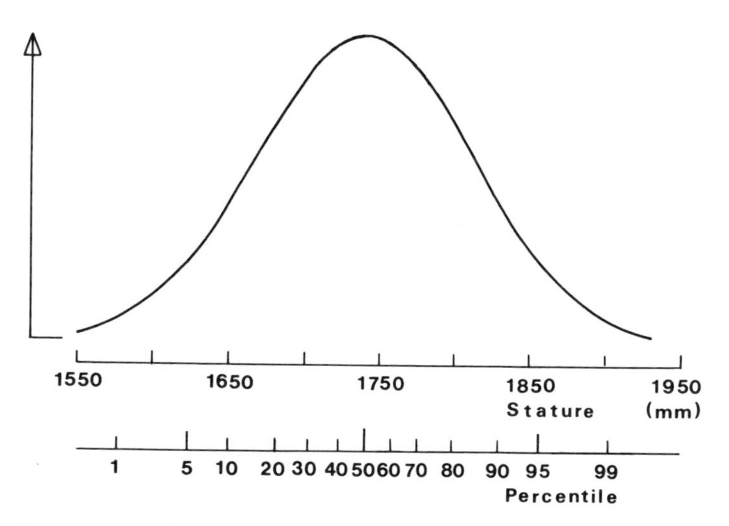

Figure 2.1. The frequency distribution (or probability density function) for the stature of adult British men. This is an example of the normal or Gaussian distribution. Data from Knight (1984).

known as the mean, is also the most probable stature. Since the curve is symmetrical, it follows that 50% of the population are shorter than average and 50% are taller—we would say, therefore, that in this distribution the mean is equal to the 50th percentile (50th %ile). In general, *n*% of people are shorter than the *n*th %ile. Hence, somewhere near the left-hand end of the horizontal axis there is a point, known as the 5th percentile (5th %ile) of which we could say "exactly 5% of people are shorter than this" or "there is only a one in twenty chance of encountering a person shorter than this". Similarly, an equal distance from the mean towards the right of the chart is a point known as the 95th %ile of which we could say "only 5% of people are taller than this". Ninety per cent of the population are between the 5th and 95th %ile in stature—but the same could be said for the 2nd and 92nd or the 3rd and 93rd. It is important to note that, by virtue of their symmetrical positions about the mean, the 5th and 95th %iles define the shortest distance along the horizontal axis to enclose 90% of the population. Two further points must be borne in mind when discussing percentiles. Firstly, percentiles are specific to the populations which they describe—hence, the 95th %ile stature for the general public might only be the 70th %ile for a specially selected occupational group like the police force or perhaps the 5th %ile for a sample made up from the Harlem Globetrotters and other professional basketball teams. Secondly, percentiles are specific to the dimension which they describe—hence, a person who is a particular percentile in stature may or may not be the same in shoulder breadth or waist circumference, since people differ in shape as well as in size.

The frequency distribution shown in Figure 2.1, with its characteristic symmetrical bell-shaped curve, is very common in biology in general and anthropometry in particular. It is usually known as the normal distribution. We should not, however, infer from this name that the distribution is in some way associated with 'normal people' as against 'abnormal' ones. We might conveniently think of the term as meaning something like "the distribution which you will find most useful in practical affairs". To avoid this possibility of confusion, some statisticians prefer to call it the 'Gaussian distribution'—after the German mathematician and physicist Johann Gauss (1777–1855) who first described it (in the context of random errors in the measurement of physical quantities). It is possible to predict that a variable such as stature will be normally distributed in the general population if we are prepared to make certain plausible assumptions concerning the way it is inherited from one generation to the next (see any textbook of genetics). Indeed, it is empirically true that most anthropometric variables conform quite closely to the normal distribution (at least within reasonably homogeneous populations). This is an exceedingly convenient state of affairs since the normal distribution may be described by a relatively simple mathematical equation. The exact form of this equation need not concern us here since we are unlikely to employ it in practice. The important thing is that it has only two parameters. (In mathematics a parameter is a quantity which is constant in the case considered but variable in different cases.) One of these parameters is the mean—it tells us where the distribution is located on the horizontal axis. The other is a quantity known as the standard deviation (s) which is an index of the degree of variability in the population concerned, i.e., the 'width' of the distribution or the extent to which individual values are scattered about or deviate from the mean. If we were to compare, for example, the general male population with the police force, we would find that the latter had a greater mean but a smaller standard deviation, i.e., they are on average taller than the rest of us and they are less variable amongst themselves. The standard deviation (s) of a sample of individuals drawn from a population is given by the equation

$$s = \sqrt{\frac{\Sigma (x - m)^2}{n - 1}} \qquad (2.1)$$

where m is the mean, x is any individual value of the dimension concerned and n is the number of subjects in the sample. (We use $n - 1$ in the equation in the hope of correcting any bias introduced by the finite size of our sample and making a better prediction of the standard deviation of the population from which it was drawn—since this is what in general concerns us.)

A normal distribution is fully defined by its mean and standard deviation—if these are known any percentile may be calculated without further reference to the raw data (i.e., the original measurements of individual people). The pth %ile of a variable is given by

$$X_{(p)} = m + sz \qquad (2.2)$$

where z is a constant for the percentile concerned, which we look up in a statistical table. A set of values for p and z are set out in Table 2.1. Suppose we wish to calculate the 90th %ile of stature for the adult male population of Britain. It happens that British men have a mean stature of 1740 mm with a standard deviation of 70 mm (see Table 4.1). From Table 2.1 we see that for $p = 90$, $z = 1{\cdot}28$. Therefore the 90th %ile value of stature $= 1740 + 70 \times 1{\cdot}28 = 1824$ mm. Alternatively, we might wish to do the calculation in reverse, and determine the percentile value for a particular stature. Hence, a stature of 1625 mm is 1·64 standard deviations below the mean. That is $z = -1{\cdot}64$. Looking this up in Table 2.1 we find that this is equivalent to the 5th %ile.

In this book we shall, in the interests of brevity, commonly adopt a convention for describing the parameters of normal distributions. Whenever a figure is followed by another in square brackets [ ] it refers to a mean and standard deviation. Hence, the statement that "the stature of British men is 1740 [70] mm" should be taken as meaning "the stature of British men is normally distributed, with a mean of 1740 mm and a standard deviation of 70 mm". (This is a purely local convention, you will not encounter it outside this book.)

Most linear dimensions of the body are normally distributed and this certainly makes life easier for the user of anthropometric data. There are, however, other

Table 2.1. p *and* z *values of the normal distribution.*

p	*z*	*p*	*z*	*p*	*z*	*p*	*z*
1	−2·33	26	−0·64	51	0·03	76	0·71
2	−2·05	27	−0·61	52	0·05	77	0·74
3	−1·88	28	−0·58	53	0·08	78	0·77
4	−1·75	29	−0·55	54	0·10	79	0·81
5	−1·64	30	−0·52	55	0·13	80	0·84
6	−1·55	31	−0·50	56	0·15	81	0·88
7	−1·48	32	−0·47	57	0·18	82	0·92
8	−1·41	33	−0·44	58	0·20	83	0·95
9	−1·34	34	−0·41	59	0·23	84	0·99
10	−1·28	35	−0·39	60	0·25	85	1·04
11	−1·23	36	−0·36	61	0·28	86	1·08
12	−1·18	37	−0·33	62	0·31	87	1·13
13	−1·13	38	−0·31	63	0·33	88	1·18
14	−1·08	39	−0·28	64	0·36	89	1·23
15	−1·04	40	−0·25	65	0·39	90	1·28
16	−0·99	41	−0·23	66	0·41	91	1·34
17	−0·95	42	−0·20	67	0·44	92	1·41
18	−0·92	43	−0·18	68	0·47	93	1·48
19	−0·88	44	−0·15	69	0·50	94	1·55
20	−0·84	45	−0·13	70	0·52	95	1·64
21	−0·81	46	−0·10	71	0·55	96	1·75
22	−0·77	47	−0·08	72	0·58	97	1·88
23	−0·74	48	−0·05	73	0·61	98	2·05
24	−0·71	49	−0·03	74	0·64	99	2·33
25	−0·67	50	0	75	0·67		

p	*z*	*p*	*z*
2·5	−1·96	97·5	1·96
0·5	−2·58	99·5	2·58
0·1	−3·09	99·9	3·09
0·01	−3·72	99·99	3·72
0·001	−4·26	99·999	4·26

kinds of frequency distribution which turn up occasionally in anthropometric practice. Some other possibilities are shown in Figure 2.2. In most populations body weight and muscular strength show a modest positive skew—it seems that there are a disproportionate number of heavy strong people and a dearth of light weak ones. Furthermore, the combination of two normal distributions, such as men and women or adults and children, will give us a new distribution that is flat-topped (platykurtic) or even double-peaked (bimodal). What will happen if we work on the erroneous assumption that such distributions are normal and go ahead and calculate percentiles by the means described above? Errors will accrue, the magnitude of which will be determined by the extent of the deviation from normality in the population distribution. The errors will in many circumstances be

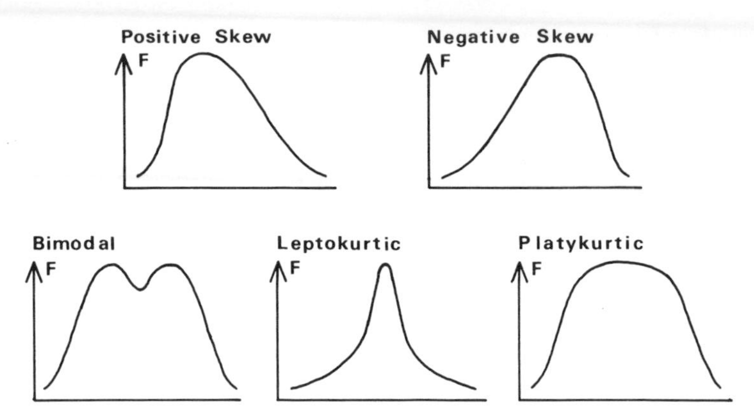

Figure 2.2. Deviations from normality in the statistical distributions of anthropometric data.

negligible. Combining data for adult men and women is a case in point. In theory, the resultant 'unisex' distribution is platykurtic. In practice, the deviations from normality are so small that we can ignore them. The only alternative, which avoids the assumption of normality, is to determine percentiles directly by simply counting heads—but since this requires large numbers of subjects it is rarely feasible and few datasets in the literature have been established with this degree of certainty. In general, the best practice is to assume normality but to proceed with circumspection in those situations (mentioned above) where we have reason to doubt the assumption. From now on, our discussion will be almost entirely confined to normal distributions.

For some purposes it may be especially informative to plot out the normal distribution in its cumulative (or integral) form. In this version percentiles are plotted against values of the dimension concerned (or, if we calibrate the horizontal axis in standard deviations, we have in effect a plot of p against z). The curve which we obtain is known as the normal ogive—see Figure 2.3. The advantage of such a plot is that, since we may read off percentiles directly, it enables us to evaluate the consequences of a design decision for the percentage of users accommodated. To take a simplistic example, Figure 2.3 would tell us directly the percentage of British men who could pass beneath an obstruction of a given height without stooping or banging their heads.

The slope of the normal ogive is greatest at the mean value (i.e., the point of maximum probability) and steadily diminishes as we approach the extreme tails of the distribution. The curve is asymptotic to the horizontal at 0 and 100% (i.e., in theory meets these lines at infinity). Hence, it is increasingly difficult to accommodate extreme percentiles. (We note in Figure 2.1 that the percentiles are densely packed near the centre and thinly spread at the extremes.) The practical consequence of this is that each successive percentage of the population we wish to accommodate imposes a more severe constraint upon our design. In cost/benefit terms we are in a condition of steadily diminishing returns. Figure 2.4 illustrates

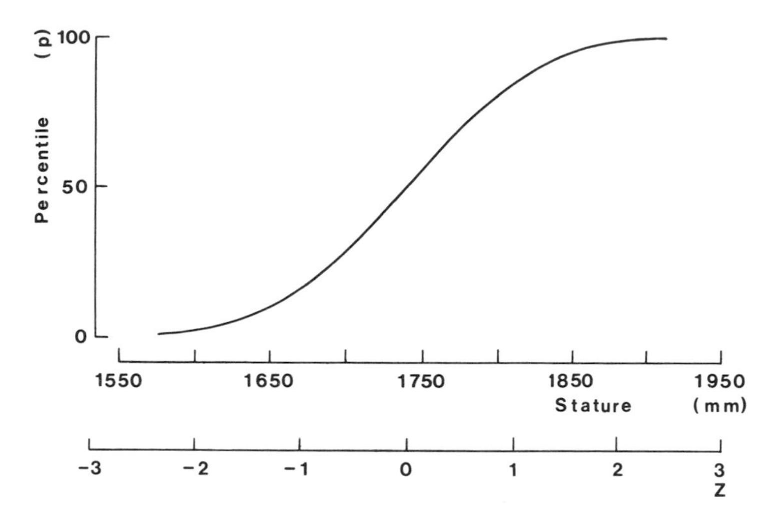

Figure 2.3. The frequency distribution of the stature of adult British men, plotted in cumulative form.

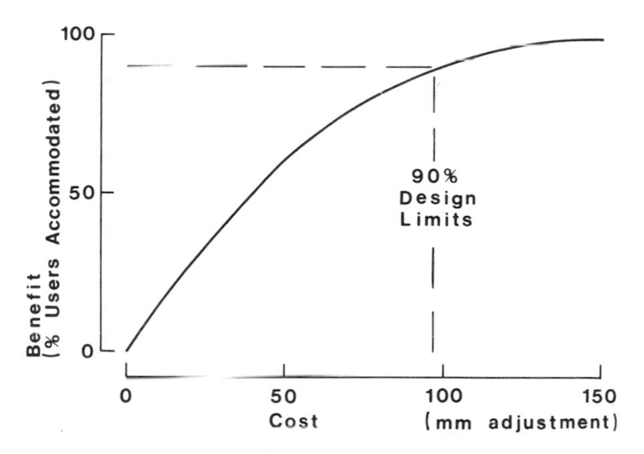

Figure 2.4. Anthropometric cost/benefit function showing the percentage of members of a target population accommodated by various ranges of adjustment in the height of a seat. The range plotted should in each case be 'centred' on a height of 455 mm.

this problem with respect to the case of the adjustability of a seat. Calculations were based on the criterion that seat height should be equal to popliteal height, which for the unisex distribution of adult British men and women (shod) = 455 [30] mm.

How then should we draw the line in this increasingly costly/constraining process of accommodating the extreme members of the user population? There is no single answer to this question. A purely arbitrary practice, which has been found satisfactory in many circumstances, is to design for the 5th–95th %ile range, i.e., the middle 90% of the population. This custom suffices well enough—but it is necessary to bear in mind the consequences of a 'mismatch' for the 10% who are outside these limits. Will it cause mild discomfort and inconvenience, or will the effective operation of the man–machine system be compromised? Are there implications for the health or safety of the user, either in the long or short term? A less than 5th %ile person sitting at a dining chair which is too high, may be mildly uncomfortable over dinner; but if she cannot easily depress the brake pedal of her motor-car, or see over the bonnet, the consequences may be more serious. The designer must exercise judgement in these matters.

In a more general sense, it is only possible to specify percentiles at all if we can first define the user population. This is simple enough if one is designing a military aircraft, for example, but the users of a public transport system would be quite another matter. Must we consider children as well as adults—or the elderly and infirm, pregnant women, or wheelchair users? These people may not fit easily into the percentile tables of the anthropometrist—but they may not be excluded from participation in the environment. The issues of barrier-free design will be addressed at a later stage (Chapter 10), but first we must consider the narrower problem of designing for the majority.

2.2. *Constraints and criteria*

In ergonomics and anthropometrics a constraint is an observable, preferably measurable, characteristic of human beings, which has consequences for the design of a particular artefact. A criterion is a standard of judgement against which the match between user and artefact may be measured. We may distinguish various hierarchic levels of criteria. Near the top are overall desiderata such as comfort, safety, efficiency, aesthetics, etc., which we may call high-level, general or primary criteria. In order to achieve these goals, numerous low-level, special or secondary criteria must be satisfied. The relationship between these concepts may be illustrated by way of example. In the design of a chair, comfort would be an obvious primary criterion; the lower leg length of the user imposes a constraint upon the design since, if the chair is too high, pressure on the underside of the thigh will cause discomfort. This leads us to propose a secondary criterion: that the seat height must not be greater than the vertical distance from the sole of the foot to the crook of the knee (this dimension is called popliteal height). A table of data will tell us the distribution of this dimension. It would seem reasonable to choose the 5th %ile value—since if a person this short in the leg is accommodated, the 95% of the population who are longer legged will also be accommodated. This leads more or less directly to a design specification or tertiary criterion: that the height of the seat shall not be greater than 400 mm. (Note that if we propose an adjustable seat we will use our criterion differently, as in Figure 2.4—see Chapter 11 for a more general discussion of this particular problem.)

Taken in isolation, the primary criterion will usually be what is known, amongst certain ergonomists of my acquaintance, as a "stunning glimpse of the obvious" (SGO). In general, it is necessary to work down through successive levels of the hierarchy before any operationally useful recommendations result. Some theorists like to contrast the 'top down' approach of working from the general to the specific with the 'bottom up' approach of working from the specific to the general.

At any level in the hierarchy conflicts between criteria may arise which will necessitate trade-offs. Hence, in the example we took above, our secondary criterion tells us when a seat is too high but not when it is too low. The criteria for this latter case are less well defined—we might call them fuzzy rather than sharp. None the less, it is perfectly possible that a tall man might feel uncomfortably cramped in a seat designed to accommodate the lower leg length of a 5th %ile woman, and some suitable compromise might have to be reached in the interest of the greatest comfort for the greatest number. Similarly, there might be circumstances in which it was necessary to trade-off, say, comfort against efficiency or safety. I suspect that these latter circumstances are few, but they raise the interesting point of what super-ordinate criterion could be used to measure both.

In practical matters, the middle of the hierarchy is often the best place to start (I have heard this called the 'middle out' approach). We shall therefore consider four sets of constraints which between them account for the vast majority of everyday problems in anthropometrics *per se* and, hence, a sizeable portion of ergonomics. We shall call them the four cardinal constraints of anthropometrics: clearance, reach, posture and strength.

Clearance

In designing workstations it is necessary to provide adequate headroom, elbow room, legroom, etc. Environments must provide adequate access and circulation space. Handles must provide adequate apertures for the fingers or palm. These are all clearance constraints. They are one-way constraints and determine the minimum acceptable dimension in the object. If such a dimension is chosen to accommodate a bulky member of the user population (e.g., 95th %ile in height, breadth, etc.), the remainder of the population, smaller than this, will necessarily be accommodated.

Reach

The ability to grasp and operate controls is an obvious example—as is the constraint mentioned above on the height of a seat or the ability to see over a visual obstruction. Reach constraints determine the maximum acceptable dimension of the object. They are again one-way constraints but this time are determined by a small member of the population, e.g., 5th %ile.

Posture

The ways in which the relationship between the dimensions of the workstation and those of the user may determine his posture, for better or worse, will be the subject of a subsequent chapter. The height of a working surface (either for standing or sitting) is a good example. In such a case it may be equally undesirable for the level to be either too high or too low—we are dealing, therefore, with a two-way constraint in which the large and small user is equally deserving of consideration. Postural criteria are often more elusive than those of reach and clearance.

Strength

A fourth constraint is concerned with acceptable limits of force in control operation or other manipulative tasks. Sometimes human strength imposes a one-way constraint and it is sufficient to estimate the loading acceptable to the weakest person—but in other cases this approach may have undesirable consequences for the heavy-handed user.

2.3. Varieties of anthropometric data

It is conventional to distinguish between 'static' and 'dynamic' anthropometric data. Regrettably, these terms are not exactly employed in the way that a physicist would use them, i.e., to denote the absence and presence of movement. This has led some authorities to suggest that they be dropped in favour of 'structural' and 'functional' anthropometrics. The latter terms have not found general acceptance, so we shall follow common usage and retain the former.

Static anthropometric data concern the fixed structural dimensions of the body, generally made between specified anatomical landmarks in stereotyped postures (see Section 4.3). Examples include stature, the heights of the eye or elbow in the standing or sitting position, the lengths of the limbs, the breadths of the shoulders or hips and the depths of the body (from front to back) at various levels. Circumferences of the limbs and trunk and body weight are also in this category.

Dynamic anthropometric data include measurements of reach or clearance made under 'functional' conditions, e.g., allowing the subjects a certain degree of freedom to adopt 'natural' postures for the performance of a given task. The range of movement of joints and the strength of various actions may also be included in this category. The direct and immediate relevance of such data to practical design problems increases with the extent to which measurement conditions approach those of the real world. Unfortunately, the price of such relevance is a high degree of situational specificity. Hence, reach measurements made in an aircraft cockpit might be irrelevant to motor-cars—by virtue of differences in seat and harness design as well as the differences in user population. As a consequence, the gathering of dynamic data must often be a 'one-off' for a particular design problem—this is expensive in time and personnel. In many cases the deficiencies of static data are not as great as they seem—they may commonly be overcome by the judicious use of criteria.

There are other types of anthropometric data which will only be mentioned in passing in this book, since they are not relevant to design problems. These include important measures of body composition such as skinfold thickness—an index of the quantity of subcutaneous fat. Similarly, the classification of physique by somatotyping will be passed over in the interests of brevity.

2.4. *Fitting trials and the method of limits*

Let us now take some of the concepts introduced above and apply them to a sample design problem. We shall analyse this problem in considerable detail—probably more detail in fact than would commonly be necessary in practice. Our task is to specify the correct working-surface height for a certain industrial assembly operation involving a modest degree of both force and precision. We may assume that, for other adequate reasons, this task must be performed standing rather than sitting, and we know that our workforce will be a representative cross-section of the adult male population. How do we commence?

One very good way would be to abandon theory altogether and take a completely empirical approach by conducting a fitting trial. For this we need an adjustable height surface on which the assembly task (or a similar activity) can be performed and a sample of subjects who are representative of the user population. Each subject would perform the task with the bench set at different heights, and make a judgement as to whether the height concerned was 'too high', 'too low' or 'just right'. We might choose to refine these judgements by having intermediate scale values. We would certainly have to take precautions not to bias our subjects' judgements—the order in which the various heights were presented would be carefully thought out with this in mind. A fitting trial is essentially a psychophysical experiment, in which observers make judgements about the sensations they experience (e.g., comfort) in response to certain physical stimuli (e.g., work-surface height). The uses of psychophysical methods in ergonomics research have been discussed by Chapanis (1959) and by Pepermans and Corlett (1983); fitting trials have been discussed by Jones (1963, 1969). Having conducted our fitting trial, we shall have, at the end of the day, a dataset which will enable us to predict the percentage of users who will find a given height satisfactory. The data will reflect not only the anthropometry variability of the subjects but also their collective experience of performing such tasks and their capacity to judge what working positions are most appropriate.

Must we, therefore, conduct a systematic fitting trial for each design problem we undertake? Such a policy is impractical for reasons of labour and expense—except for special occasions. It is also unnecessary since, on many occasions, we may achieve comparable results by pencil-and-paper analyses. We may choose to view such techniques as a model or analogue of the fitting trial in which anthropometric data and criteria are used to represent real people—as a substitute so to speak.

Grandjean (1981) recommends that the best level for performing manipulative tasks of moderate force and precision is between 50 and 100 mm below the height of the elbow (see Section 8.5). This will be our criterion. We note that it is a two-way criterion since it may be exceeded in either direction. The elbow height (EH) of British men is 1090 [52] mm (see Table 4.1). To this we must add a 25 mm correction for shoes giving 1115 [52] mm (see Section 4.2). Combining these data with the above criterion gives us the upper and lower limits of optimal working level. EH − 50 = 1065 [52]; EH − 100 = 1015 [52]. We can treat these just as if they are new normally distributed anthropometric dimensions—and calculate the percentile in these distributions to which any particular workbench height corresponds. However, we should bear in mind that the criteria refer to 'optimal' bench heights. Since we may reasonably assume that users may be prepared to accept less than absolute perfection, we may well find it useful to consider two further zones extending 50 mm above and below the optimum, which we would characterize as 'satisfactory but not perfect' (see Figure 2.5). We choose the figure of 50 mm because it 'seems reasonable' rather than on the basis of any particular scientific evidence.

Table 2.2 shows a complete set of calculations performed for the height of 1000 mm. We find that this corresponds to the 75th %ile in the lowest criterion distribution—from which we infer that a workbench of 1000 mm would be 'much too low', or 'unsatisfactory', for the 25% of men who are taller than this. Similarly,

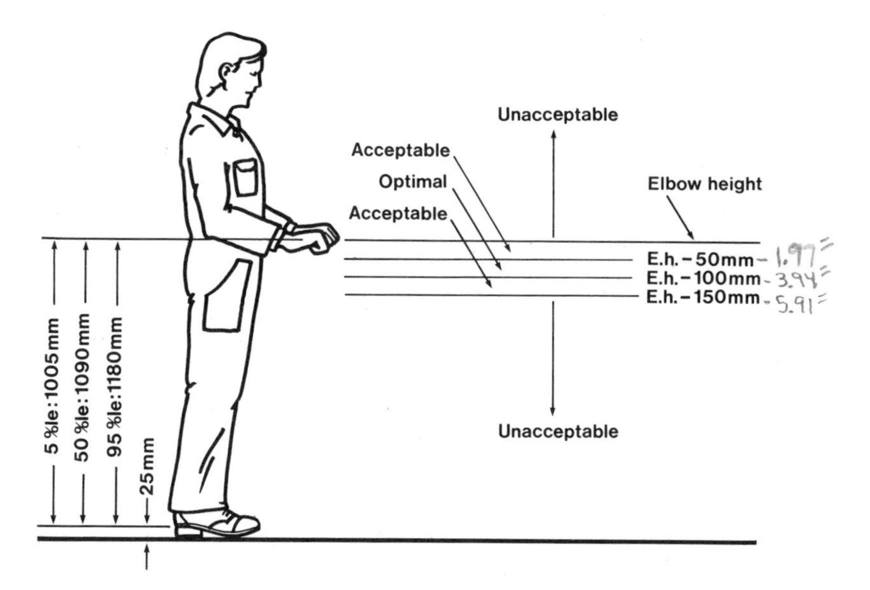

Figure 2.5. Criteria for optimal and satisfactory working heights in an industrial assembly task.

Table 2.2. Calculation of percentage of men accommodated by a workbench which is 1000 mm in height.

Criterion	Distribution	Percentile	Conclusion
EH − 150	965 [52]	75	25%—much too low
EH − 100	1015 [52]	39	61%—too low
EH − 50	1065 [52]	11	11%—too high
EH	1115 [52]	1	1%—much too high
			28%—just right

the centre criteria correspond to the 39th and 11th %iles, respectively—from which we infer that the 28% of men between these heights would find the workbench 'just right' or 'optimal'.

We could keep on performing such calculations for different heights until we homed in on a value which maximized the percentage optimally matched and minimized the unsatisfactory matches. (Here, of course, the computer would help.) At this point we are like the statistically minded punter searching for the best bet. The results of a series of such calculations are plotted in Figure 2.6. It comes as no surprise to discover that the 'optimal' figures describe a normal curve (e); whereas the 'too high' and 'too low' figures yield normal ogives facing in opposite directions (a, b, c, d). We might also lump together those who were optimally matched with those who were a little too high and a little too low into a 'satisfactory' category (f), leaving a residual 'unsatisfactory' figure (g) outside these limits (26% unsatisfactory and 74% satisfactory for 1000 mm). The statistically minded punter should settle for a working height of 1050 mm.

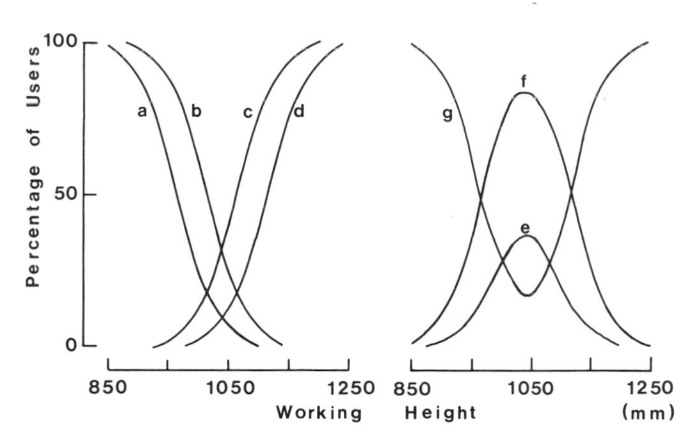

Figure 2.6. The anthropometric method of limits, applied to the determination of the optimal working height, for an industrial assembly task. Curves show the percentage of users accommodated or otherwise: (a) much too low; (b) too low; (c) too high; (d) much too high; (e) just right; (f) satisfactory; (g) unsatisfactory. See text for definition of categories and discussion of technique.

This is not quite the end of the process since at the chosen figure some 15% of users will have an 'unsatisfactory' match. Is this an acceptable or tolerable situation or will they be severely uncomfortable or suffer long-term damage? Is it better to have a bench which is too high or one which is too low? Do we in fact require an adjustable workbench or some similarly varied solution? Partial answers to these questions will be found in Chapter 8. There will be some situations in which the only way to solve problems of this kind will be by testing out a prototype—this may seem to take us full circle to the 'fitting trial approach', but the statistical analysis has narrowed our range of options to a point where this is more feasible. Anthropometrics remains a relatively inexact science and most ergonomists would consider a user trial at the prototype stage to be an essential phase of the design process (in the same way as engineers test models in wind tunnels to check their calculations).

The procedure we have adopted in searching for the best compromise is sometimes known as the method of limits. This name, which is borrowed from that of an experimental technique in psychophysics which is broadly similar in form, reinforces the fact that it is in essence an imaginary fitting trial.

2.5. *Variation in shape and bodily proportion*

It is a matter of common experience that human beings, even when sampled from a relatively homogeneous population in terms of age, sex and ethnicity, vary considerably in shape as well as in overall size. If we were to select a sample of men of similar stature, we would expect to see considerable variation in their bodily proportions. The practical consequences of this very elementary observation are very considerable.

Anthropologists have traditionally sought to describe the proportions of the body by the ratios of one dimension to another, usually expressed as percentage indices. Hence,

$$\text{Sitting height index} = \frac{\text{sitting height}}{\text{stature}} \times 100\% \qquad (2.3)$$

This is a description of an individual, and as such it is distributed within a population and we may calculate its mean and standard deviation in the usual way. In an extensive survey of US civilians (Stoudt *et al.* 1965, 1970) the sitting height index was 51·8 [1·5] for men and 52·4 [1·5] for women (but the distribution was slightly skewed). The distribution of the sitting height index must be calculated from the original raw data of the survey (unless we use some very advanced statistical techniques). When these are not available (as is generally the case) we may use, as a substitute, a relative dimension. Hence,

$$\text{Relative sitting height} = \frac{\text{mean sitting height}}{\text{mean stature}} \times 100\% \qquad (2.4)$$

For reasons which would only be apparent to a mathematical statistician mean (x/y) is not equal to mean x/y, hence the relative sitting height is an approximation to the mean sitting height index but is not numerically identical. (The terms we have just defined are sometimes used interchangeably in the literature—but they will be used consistently in the present book.)

Another way of looking at two variables at once, which is statistically more sophisticated, is through the concept of a bivariate normal distribution and the techniques of correlation and regression. A mathematical treatment of these topics will be found in any statistical textbook and the equations involved are summarized in Section 2.7. For the moment we shall consider the matter qualitatively. Consider two normally distributed variables—shoulder breadth and stature, for example. The degree of relatedness or association between these variables may be expressed by a parameter called the correlation coefficient (r). If the two variables are totally unrelated $r = 0$; if they perfectly relate in a totally predictable way $r = \pm 1$ ($+1$ if directly proportional, -1 if inversely). A bivariate normal distribution is completely defined by the mean and standard deviation of the two variables concerned and their correlation coefficient. The probability function for a bivariate normal distribution is a three-dimensional curved surface. Rather than trying to draw such surfaces in projection, it is simpler to represent them as contour maps on which the lines indicate equal probability—as shown in Figure 2.7. When

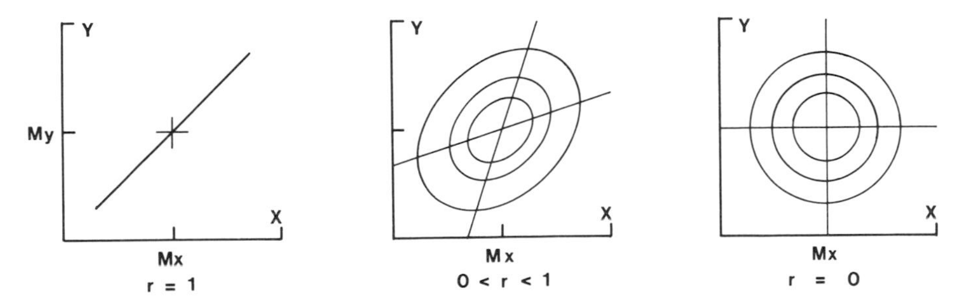

Figure 2.7. Equal probability contours of bivariate normal distributions with different levels of correlation. M_x *and* M_y *are the mean values of the variables* x *and* y.

variables x and y are uncorrelated ($r = 0$), equal probability contours are concentric circles—the surface they represent would be the bell shape which is described by rotating the normal curve through 360°. As r increases, this shape is gradually flattened and the circles become ellipses. When the variables are perfectly correlated ($r = 1$) the surface has been flattened into a plane and the ellipses have become a straight line.

Several points of interest emerge from these considerations. The first concerns 'the fallacy of the average man (person)'. Supposing, as an arbitrary (somewhat generous) definition of 'average', we say that a person should be in the middle quarter of the distribution of a given variable. The chances of being similarly average in three uncorrelated variables are $\frac{1}{4} \times \frac{1}{4} \times \frac{1}{4}$ or 1 in 64 (i.e., less than 2%). Now, of course, anthropometric variables are correlated and some pairs are strongly correlated. None the less, the chances of encountering an individual who is 'average' in a number of dimensions are extremely small. But before we rush to conclude that 'the average man does not exist' we should bear in mind that he remains the most probable member of the population. The previous paragraphs described a bivariate normal distribution—it is more difficult to visualize a multivariate distribution since its probability function can only be drawn in hyperspace. None the less, that function will be maximal when all dimensions are average. The art of stocking a store with 'off the peg' clothing rests on an informed judgement as to what combinations of dimension are more probable than others.

More importantly, it is mathematically impossible for an individual to be, say, 95th %ile in all respects. Consider, for example, the consequences of adding the 95th %ile trunk length to the 95th %ile lower limb length. The result must necessarily be somewhat greater than the 95th %ile stature—since some tall people have disproportionally long legs and others disproportionally long trunks (i.e., since we are not all the same shape the correlation between trunk length or leg length and stature is less than 1). Design methods which assume a '95th %ile person' or a '5th %ile person' are based on convenient fictions. Unfortunately, alternative and mathematically more rigorous approaches may be dauntingly complex—in general it is sufficient that the designer should be aware of the limitations of simplistic assumptions concerning percentiles and take care to ensure that these do not lead him to commit major errors.

Supposing a workstation, such as a control console, was designed to take into account several different anthropometric constraints (knee height, sitting elbow height, sitting eye height, etc.) and at each stage a dimension which gave 95% accommodation was chosen—at the end of the day what percentage of users would be accommodated? Each successive dimension excludes 5% of users—but since the dimensions are not perfectly correlated a different 5% are excluded each time. In principle, this exclusion could mount up alarmingly! Bittner *et al.* (1975) studied the matter by a computer technique called 'Monte Carlo modelling'—in essence, the computer generates a string of random numbers which have the same multivariate normal distribution as the anthropometric data of the subjects. (We need not concern ourselves with how this prodigious feat is achieved.) The outcome of the calculations is shown in Figure 2.8. The intended (nominal) accommodation is plotted horizontally; the actual percentage accommodated when seven separate anthropometric constraints are considered together is plotted vertically. It is sobering to realize that if we rigorously follow a policy of accommodating 95% of the population (i.e., excluding 5%) we actually end up excluding 25%.

2.6. *Mannikins, models and computers*

Since many of us find the abstract world of statistical tables somewhat uninviting, a considerable demand exists for anthropometric data that have been 'compressed'

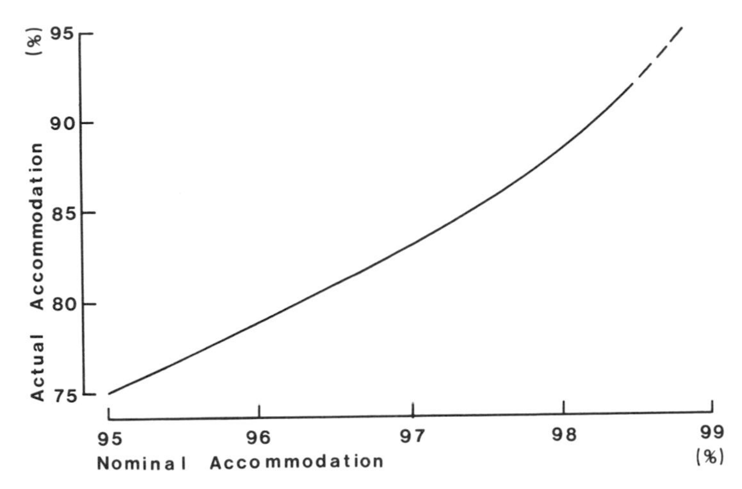

Figure 2.8. Nominal and actual accommodation in a workspace design involving seven constraints as calculated by Monte Carlo modelling.
Data from Bittner et al. *(1975).*

into a more readily usable form—in fact, for a model of the data which can be manipulated so that the consequences of design decisions may be readily evaluated, if possible, by direct visual means. The simplest of these models is a two-dimensional drawing board mannikin which articulates in a way that (approximately) resembles the human body.

It is worth noting that, in theory at least, a 95th %ile mannikin is an impossibility for reasons given in the last section. None the less, a sensible compromise may be achieved in which the most critical dimensions are preserved. Furthermore, some commercially available drawing aids seemingly represent elegant idealized human beings rather than any percentile in particular.

The mannikin drawings in Figures 2.9–2.12 are based on the data of the present book (as given in Table 4.1 and elsewhere) and I believe them to be as accurate as is reasonably practicable. Life-size versions of these are obtainable from Communications Complex Design (CCD) Ltd (76 Church Street, Weybridge, Surrey KT13 8DL, England). Mannikins for other percentiles and other ethnic groups may be obtained from the same source.

It seems likely that the immediate future will see major advances in computer modelling of anthropometric data. The computer program called SAMMIE (System for Aiding Man–Machine Interaction and Evaluation), developed by Professor Maurice Bonney and his colleagues, is a good example of the current generation of models. In essence, the computer stores a sufficiently detailed anthropometric table to generate a three-dimensional image of a user who is a specified percentile in certain respects. This image may then be displayed on the screen in elevation, plan, projection or perspective and may be moved around under operator control in a similarly represented environment, in order to evaluate

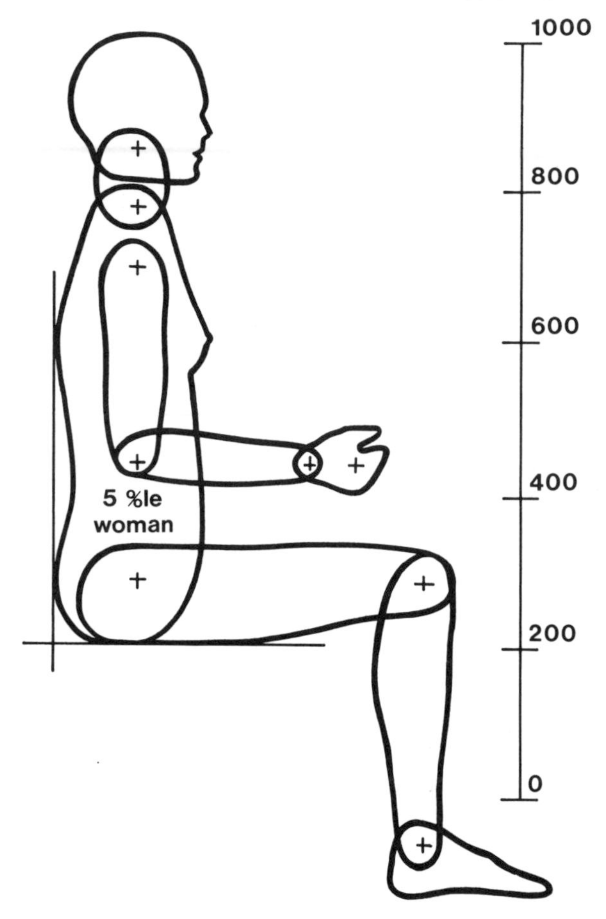

Figure 2.9. Mannikin—5th %ile woman (scale 1:10).

projected layouts with respect to reach, fit, clearance and field of view. The system can be used both as a design and an evaluation tool and has been used for design problems associated with vehicles (trucks, cars, tractors, ships, aircraft and trains) and other diverse tasks and equipment such as the design of cash payment areas in shops, the design of computer-aided equipment and the layout of control rooms. Figure 2.13 illustrates a rather unusual application. It shows a tall female mannikin (a nurse) reaching into a specially designed cot for infant care. Various dimensions of the cot were examined in terms of comfort and ease-of-use criteria. Recommendations based on these SAMMIE evaluations were then fed back to the designer prior to user trials. (This particular project was performed by Stuart Gibson and Ian Pemberton).

An alternative type of program is one like the Monte Carlo model of Bittner *et al.* (1975) known as CAPE (Computerized Accommodated Percentage

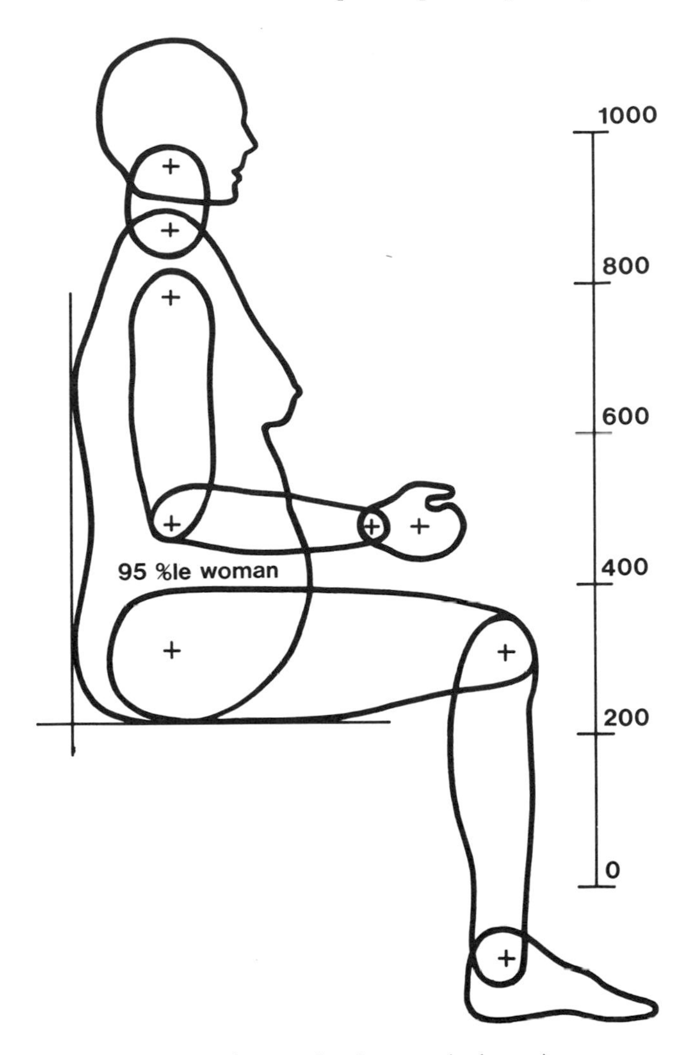

Figure 2.10. Mannikin—95th %ile woman (scale 1:10).

Evaluation), the nature and application of which was described previously.

It is difficult to predict the effects the increasing availability of such software will have on the practice of anthropometrics and design. Consider the possibility of a software package which incorporated an extensive anthropometric data library, with a fixed percentile model like SAMMIE and a percentage accommodated system like CAPE, which was fully interactive and had some of the features of an 'expert' or 'intelligent knowledge-based system', which produced presentation quality graphics and which ran on a cheap desk-top micro. Such a package would be a very powerful design tool—by the time you are reading this it may well be available.

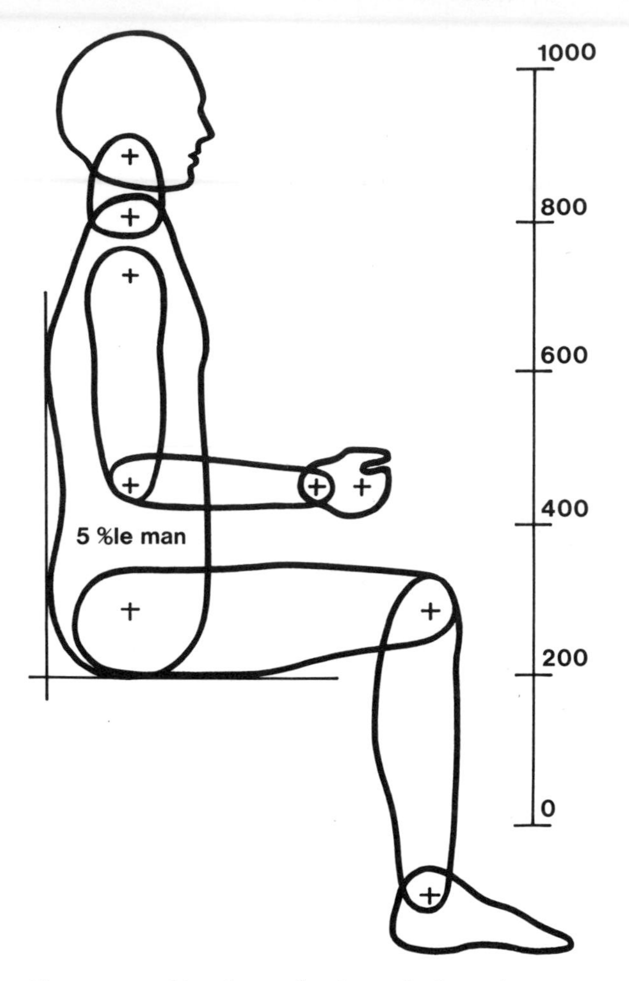

Figure 2.11. Mannikin—5th %ile man (scale 1:10).

2.7. *A mathematical synopsis of anthropometrics*

The purpose of this section is to treat the mathematical foundations of anthropometrics in a little more detail than hitherto. The reader who has no taste for these matters may skip this section with impunity since the sense of the remainder of the book does not depend on it to any critical extent. The equations proposed here will, however, be used extensively.

The normal distribution

Consider a variable x which is normally distributed in a population. Its probability density function is

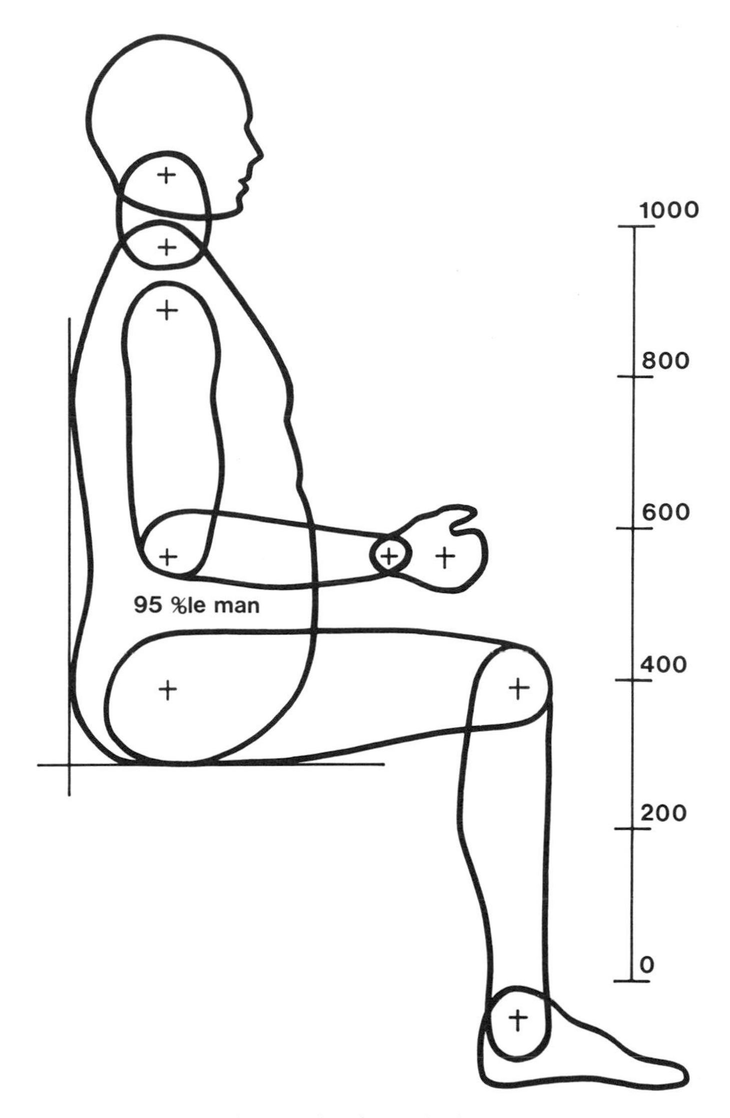

Figure 2.12. *Mannikin—95th %ile man (scale 1:10).*

$$f(x) = (1/\sigma\sqrt{2\pi})\ \exp\ [-(x-\mu)^2/2\sigma^2] \tag{2.5}$$

where μ is the mean and σ is the standard deviation of x for the population. $f(x)$ is a measure of the relative probability or relative frequency of the variable having a given value of x—it will be found in books of statistical tables as the "ordinate of the normal curve". If variable x is replaced by the standard normal deviate (z), such that

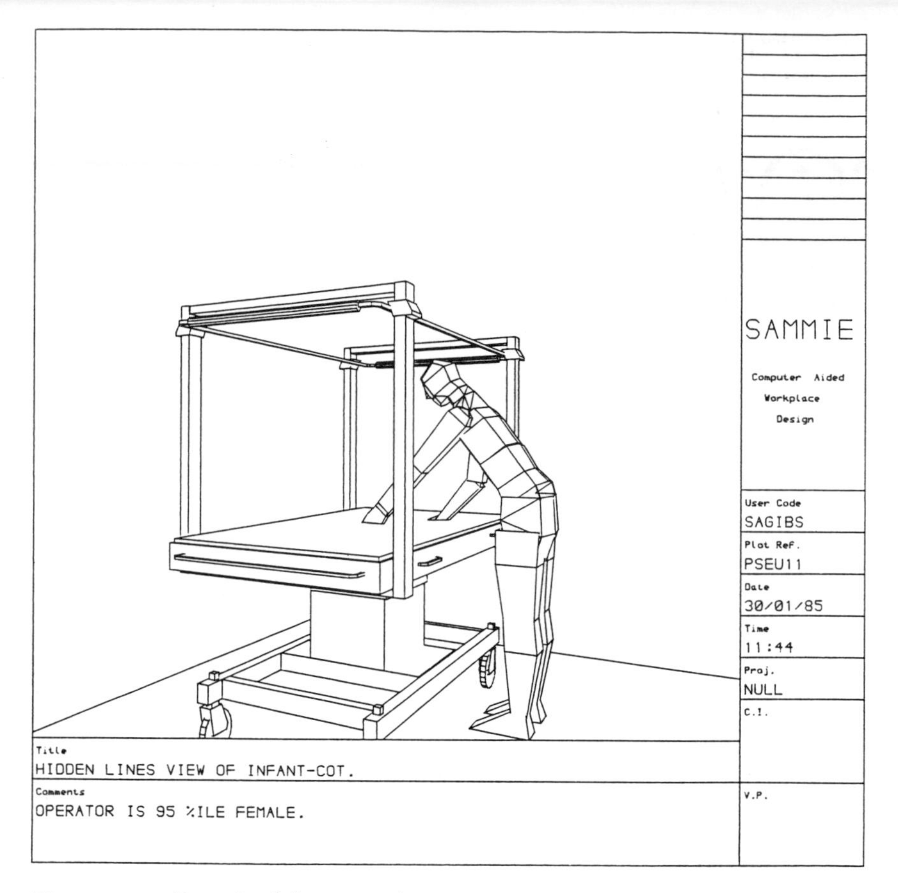

Figure 2.13. Example of the output of a computer model called SAMMIE (supplied by Professor M. Bonney).

$$z = (x - \mu)/\sigma \tag{2.6}$$

Equation 2.5 becomes

$$f(z) = (1/\sqrt{2\pi}) \exp(-z^2/2) \tag{2.7}$$

which is known as the standardized form of the normal distribution (with zero mean and unit standard deviation).

The probability that x is less than or equal to a certain value is given by

$$F(x) = \int_{-\infty}^{x} f(x)\, dx \tag{2.8}$$

i.e., $F(x)$ corresponds to the area between the abscissa and the curve from $-\infty$ to x. This is the cumulative normal curve or normal ogive. It is tabulated in Table 2.1 in which $F(x)$ is given as a percentage (p) for given values of z (as defined above).

Samples, populations and errors

In reality we cannot know μ and σ, the parameters of a population (except in very special circumstances). We can only infer or estimate them from m and s—the mean and standard deviation of a sample of individuals deemed to be representative of the population, such that

$$m = \Sigma\, x/n \tag{2.9}$$

and

$$s = \sqrt{\Sigma\,(x - m)^2/n} \tag{2.10}$$

where n is the number of subjects in the sample.

In small samples (e.g., $n = 30$) it is conventional to make the arbitrary correction

$$s = \sqrt{\Sigma\,(x - m)^2/(n - 1)} \tag{2.11}$$

The quantity s is known as the variance of the sample.

In many cases it is more convenient to calculate the standard deviation by means of the identity

$$\Sigma(x - m)^2 = \Sigma x^2 - (\Sigma x)^2/n \tag{2.12}$$

As n increases, m and s become more reliable estimates of μ and σ, i.e., the likely magnitude of random sampling errors diminishes. (Note that we are not talking about errors of bias due to non-representative sampling—this is a more complex matter.) Sampling errors in estimating population parameters may be shown to be normally distributed with a mean of zero and a standard deviation known as the standard error (SE) of the parameter concerned, such that

$$\text{SE mean} = s/\sqrt{n} \tag{2.13}$$

$$\text{SE standard deviation} = s/\sqrt{2n} = 0{\cdot}71\ \text{SE mean} \tag{2.14}$$

$$SE\ p\text{th \%ile} = \frac{p(100 - p)s}{100\, f_p n} \tag{2.15}$$

where f_p is the ordinate of the normal curve at the pth %ile.

Probable magnitudes of sampling errors are commonly expressed in terms of the 95% confidence limits of the parameter concerned, which are calculated as ±1·96 SE, i.e., the true values of a population parameter will lie within ±1·96 standard errors of the sample statistic, 95 times out of every 100 that the sample is drawn. (Alternatively, if we are concerned with errors in one direction only, we use 1·645 SE.)

To simplify matters we may summarize this by saying that in any anthropometric survey the 95% confidence limits of a statistic ($\pm U_{95}$) are given by

$$U_{95} = ks/\sqrt{n} \qquad (2.16)$$

where k is a constant for the statistic concerned given in Table 2.3, or, alternatively, the equation

$$n = \left(\frac{ks}{U_{95}}\right)^2 \qquad (2.17)$$

gives us an indication of the number of subjects we need to measure in order for a particular statistic to give a certain desired degree of accuracy.

Table 2.3. Values of the parameter k, *as given in equation 2.16.*

Statistic	k
Mean	1·96
Standard deviation	1·39
Percentiles	
40th and 60th	2·49
30th and 70th	2·58
20th and 80th	2·80
10th and 90th	3·35
5th and 95th	4·14
1st and 99th	7·33

The coefficient of variation

The coefficient of variation (CV) is given by

$$\mathrm{CV} = s/m \times 100\% \qquad (2.18)$$

It is a useful index of the inherent variability of a dimension, i.e., it is independent both of absolute magnitude and of units of measurement.

In most populations stature has a lower CV than any other dimension. (Does this reflect a biological phenomenon or is it an artefact of measurement?) Characteristic ranges of CV of various types of anthropometric data are shown in

Table 2.4. Characteristic coefficients of variation of anthropometric data.

Dimension	CV (%)
Stature	3–4
Body heights (sitting height, elbow height, etc.)	3–5
Parts of limbs	4–5
Body breadths (hip, shoulder, etc.)	5–9
Body depths (abdominal, chest, etc.)	6–9
Dynamic reach	4–11
Weight	10–21
Joint ranges	7–28
Muscular strength (static)	13–85

Table 2.4. The figures were gathered from a number of sources (Damon *et al.* 1966, Roebuck *et al.* 1975, Grieve and Pheasant 1982) and do not reflect any specific population. They should rather be seen as a general guide to the approximate levels which we might anticipate. The high CVs of the lower part of the table are indicative of a skewed distribution—which is characteristic of anthropometric dimensions including soft tissue (fat) and of functional measures such as strength.

Roebuck *et al.* (1975) have demonstrated that for body length and breadth dimensions, in general, the relationship between standard deviation and mean will tend to be curvilinear (i.e., CV declines with increasing mean value). The reasons behind this observation are obscure but may be concerned with measurement error. Figure 2.14 shows this relationship plotted out for the 36 dimensions of Table 4.1.

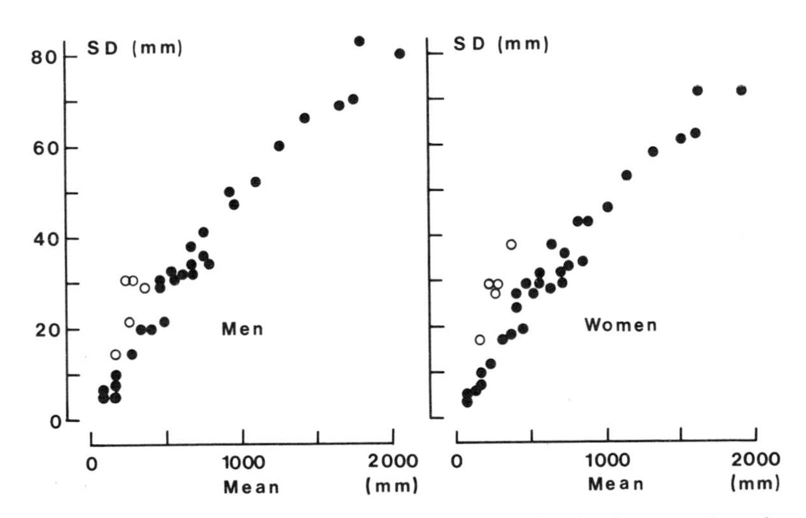

Figure 2.14. Relationship between standard deviation (SD) and mean in the anthropometric data of Table 4.1.
○ = *body depths, thigh thickness, sitting elbow height and hip breadth;* ● = *all other dimensions.*

Combining distributions

There are numerous situations in which the parameters of two or more normally distributed samples or populations must be combined to give a single lumped distribution. Strictly speaking the new lumped distribution cannot be normal (Gaussian). To describe it we should calculate percentiles iteratively. Consider two sample distributions $m_1[s_1]$ and $m_2[s_2]$ of n_1 and n_2 subjects, respectively. For any value of x calculate standard normal deviates z_1 and z_2 in the two distributions and convert to percentiles p_1 and p_2 using Table 2.1. The percentile p' in the lumped distributions is given by

$$p' = \frac{p_1 n_1 + p_2 n_2}{n_1 + n_2} \tag{2.19}$$

To describe the lumped distribution this process is repeated iteratively for as many values of x as is required.

However, in many cases the lumped distribution may be approximated by a new normal distribution $m'[s']$. In order to do this we must recalculate the sum $(\Sigma\, x)$ and the sum of squares $(\Sigma\, x^2)$ of the original raw data. For each normal distribution

$$\Sigma\, x = nm \tag{2.20}$$

and from equations 2.10 and 2.12

$$\Sigma\, x^2 = ns^2 + nm^2 \tag{2.21}$$

and for k samples described by m_i, s_i, n_i, etc.

$$\Sigma\, x = \sum_{i=1}^{k} (n_i m_i) \tag{2.22}$$

$$\Sigma\, x^2 = \sum_{i=1}^{k} (n_i s_i^2 + n_i m_i^2) \tag{2.23}$$

and

$$n' = \sum_{i=1}^{k} (n_i) \tag{2.24}$$

Therefore,

$$m' = \Sigma(n_i m_i)/\Sigma(n_i) \tag{2.25}$$

$$(s')^2 = \frac{\{\Sigma(n_i s_i^2 + n_i m_i^2) - [\,\Sigma\,(n_i m_i)]^2/\Sigma\, n_i\}}{n_i} \tag{2.26}$$

The validity of the latter approach is greatest when the constituent sample standard deviations are large and the differences between the means are small.

In the case when n is the same for all k samples then

$$m' = \Sigma m_i/k \tag{2.27}$$

and

$$(s')^2 = \frac{\Sigma(m_i^2 + s_i^2) - (\Sigma m_i)^2/k}{k} \tag{2.28}$$

The bivariate distribution

Consider two normally distributed variables x and y. Their joint probability density function is given by

$$f_{(xy)} = 1/2\pi\sigma_x\sigma_y\sqrt{1-\rho^2}\,e^w \tag{2.29}$$

where

$$w = -(x-\mu_x)^2/\sigma_x - 2\rho(x-\mu_x)(y-\mu_y)/\sigma_x\sigma_y + \frac{(y-\mu_y)^2\sigma_y^2}{2(1-\rho^2)} \tag{2.30}$$

This distribution has five parameters of which μ_x, μ_y, μ_x, μ_y are self-explanatory and ρ is the correlation coefficient for the population which is best estimated by the sample correlation coefficient (r), where

$$r = \frac{\Sigma(x-m_x)(y-m_y)}{\sqrt{\Sigma(x-m_x)^2\ \Sigma(y-m_y)^2}} \tag{2.31}$$

or

$$r = s_{xy}/s_x s_y \tag{2.32}$$

where

$$s_{xy} = \frac{\Sigma(x-m_x)(y-m_y)}{n} \tag{2.33}$$

which is known as the covariance of x and y and s_x and s_y are the sample standard deviations of x and y, respectively.

If the surface of the bivariate probability function is cut by a plane parallel to the y axis, the intersection will describe a normal probability curve. This curve defines the distribution of values of y as a population of subjects who all have the same value of x; and the mean of this distribution is the most probable value of y for a given value of x. The means of all such distributions fall on a straight line known as the regression line of y on x, given by the equation

$$y = a + bx \tag{2.34}$$

where

$$b = \frac{\Sigma (x - m_x)(y - m_y)}{\Sigma (x - m_x)^2} = \frac{s_{xy}}{s_x^2} = \frac{rs_y}{s_x} \tag{2.35}$$

and

$$a = m_y - bm_x \tag{2.36}$$

(Similarly, the means of normal distributions, given by sections cutting the bivariate distribution parallel to the x axis, define the regression line of x and y. The two regression lines will be co-incident if r is 1; perpendicular if r is 0.)

Equations 2.34–2.36 may also be written

$$y - m_y = r(s_x/s_y)(x - m_x) \tag{2.37}$$

At any given value of x, y is normally distributed with a mean defined by the regression line and a standard deviation given by

$$\text{SE} = s_y \sqrt{1 - r^2} \tag{2.38}$$

(SE being known as the standard error of the estimate.)

Estimating unknown distributions

The practical anthropometrist is frequently required to estimate the distribution of a dimension which, for reasons of practical expediency, may not be measured directly in a particular population. Some useful techniques will be described.

Correlation and regression

If the parameters m_x, m_y, s_x, s_y and r are known in sample 1, and the parameters m_x and s_x are known in sample 2, then m_y and s_y may be estimated in sample 2 from the equations given above on the assumption that r is the same in both samples. (These samples may, of course, be deemed to be representative of populations.)

Sum and difference dimensions

When an unknown dimension is anatomically equivalent to the sum of two known dimensions then

$$m_{(x+y)} = m_x + m_y \tag{2.39}$$

$$s_{(x+y)}^2 = s_x^2 + s_y^2 + 2rs_xs_y \tag{2.40}$$

When an unknown dimension is anatomically equivalent to the difference between two known dimensions then

$$m_{(x-y)} = m_x - m_y \quad (2.41)$$

$$s^2_{(x-y)} = s^2_x + s^2_y - 2rs_xs_y \quad (2.42)$$

The method of ratio scaling

If the parameters of variables x and y are known in a reference population a (or more precisely in a sample drawn from it) but only the parameters of x are known in population b (which we shall call the 'target population') then

$$m_y/m_x \text{ (in population a)} \simeq m_y/m_x \text{ (in population b)} \quad (2.43)$$

and

$$s_y/s_x \text{ (in population a)} \simeq s_y/s_x \text{ (in population b)} \quad (2.44)$$

provided that populations a and b are similar in age, sex and ethnicity.

Although these equations cannot be justified mathematically they have been widely employed, both in the present text and elsewhere, on grounds of practical expediency (e.g., Barkla 1961). We may call the dimension x, which is known in both populations, the scaling dimension—stature is most commonly used for this purpose since it is commonly available for populations in which other data are sparse. The simplest technique is to collect, from a variety of reference populations, the coefficients

$$E_1 = \frac{\text{mean of required dimension}}{\text{mean stature}} \quad (2.45)$$

$$E_2 = \frac{\text{standard deviation at required dimension}}{\text{standard deviation of stature}} \quad (2.46)$$

and simply multiply these by the parameters of stature in the target population. (Note that coefficient E_1 is in fact the relative value of the dimension as defined in Equation 2.4.) Pheasant (1982 a) conducted a validation study of this technique and found that its errors were acceptable for most purposes. (See Section 4.1 for a further discussion.)

Empirical estimation of standard deviation

The data plotted in Figure 2.11 may be empirically fitted with the following regression equations.

For body heights, lengths and breadths:

$$\text{men:} \quad s = 0{\cdot}05703m - 0{\cdot}000008347m^2 \qquad (2.47)$$

$$\text{women:} \quad s = 0{\cdot}05783m - 0{\cdot}000010647m^2 \qquad (2.48)$$

For body depths, thigh thickness, sitting elbow height and hip breadth:

$$\text{men:} \quad s = 7{\cdot}864 + 0{\cdot}06977m \qquad (2.49)$$

$$\text{women:} \quad s = 4{\cdot}249 + 0{\cdot}09467m \qquad (2.50)$$

These equations may then be used as a first estimate of the standard deviation of a dimension for which the mean is known or can be calculated.

Alternatively, if the coefficient of variation of a similar dimension is known, it may be assumed that the CV of the required dimension is the same and the standard deviation may again be calculated from the mean. (If Equations 2.47–2.50 hold then this latter assumption will tend to overestimate the standard deviation of large dimensions and underestimate that of small ones.)

Combinations of people

Consider distributions $m_a[s_a]$ and $m_b[s_b]$. (These might be the same variable in different samples of individuals or different variables in the same sample.)

If members of the two distributions meet at random (i.e., chance encounters occur) the distribution of differences is given by

$$m_{(a-b)} = m_a - m_b \qquad (2.51)$$

$$s^2_{(a-b)} = s^2_a + s^2_b \qquad (2.52)$$

and the distribution of sums by

$$m_{(a-b)} = m_a + m_b \qquad (2.53)$$

$$s^2_{(a+b)} = s^2_a + s^2_b \qquad (2.54)$$

(Applications of these equations are found in Sections 3.1 and 11.4.)

In certain design applications it is necessary to know the breadth of two or more people placed side by side, e.g., upon a bench seat. If the body breadth concerned has the distribution $m[s]$ then the parameters of the group distribution are given by

$$m_g = nm \tag{2.55}$$

$$s_g = s\sqrt{n} \tag{2.56}$$

where n is the number of people in the group.

Chapter 3
Human diversity

The purpose of this chapter is to analyse, as far as we are able in the light of present knowledge, the ways in which the biological and social differences between groups of human beings may affect their anthropometric characteristics. The study of these differences has led more than one ergonomist to comment that (data concerning) the proportions or strengths of human beings are commonly broken down by age and sex. Our list of 'biosocial' variables must also include ethnicity, social class and occupation. Furthermore, historical processes acting over at least a century have led to an almost worldwide steady increase in body size (referred to in the literature as 'the secular trend'). Attempting to analyse the interactions of these factors is something like trying to assemble a jigsaw puzzle when you strongly suspect that many of the pieces are missing.

The extent to which the measurable differences between groups of people are determined by biological (genetic) forces or social (environmental) ones requires consideration. This 'nature/nurture' question is often a very difficult one in the human sciences—asking whether a given characteristic is determined by inheritance or lifestyle is often like asking whether the area of a rectangle is determined by its length or its breadth.

Let us commence by considering the distribution of the variable stature in the human race as a whole (Figure 3.1). It is unlikely that the anthropometric data available at present, even supposing they could all be assembled in one place, constitute a representative sample of mankind. The data tabulated by Tildesley (1950) suggest that the average stature of all living male adults is, in round numbers, 1650 [80] mm. (The reader will recall from Section 2.1 that a figure followed by another in square brackets indicates the mean and standard deviation of a normal distribution.) Assuming a sex difference of 7% in the mean, and equal coefficients of variation, living female adults have a stature distribution in the order of 1535 [75] mm. (These figures could no doubt be improved by anyone with the patience to do so but will serve for the purpose of argument.) The adult population of Great Britain is well into the taller half of the human race: 1740 [70] mm for men and 1610 [62] mm for women. Hence, the average British male is the 87th %ile for the human race as a whole.

According to Roberts (1975) the shortest people in the world are the Efe and Basua 'pygmies' of central Africa, whose average stature is 1438 [70] mm for men

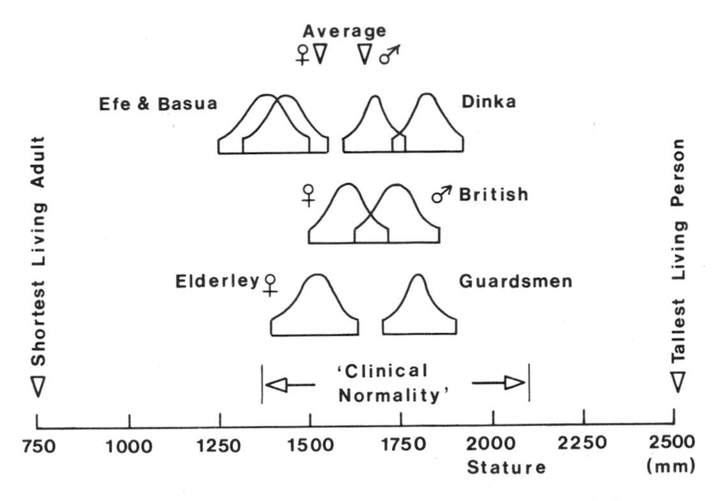

Figure 3.1. Variation in adult stature in the human race as a whole.

and 1372 [78] mm for women. The tallest are the Dinka Nilotes of the southern Sudan: 1829 [61] mm for men and 1689 [58] mm for women. However, differences almost as great as these may be found between particular samples drawn from the British population. Guardsmen (Gooderson and Beebee 1977) stand at some 1803 [63] mm whereas a sample of elderly women measured by Caroline Harris and her colleagues (Institute for Consumer Ergonomics 1983) had a stature of 1515 [70] mm (deducting a modest 20 mm for shoes since the subjects were measured shod).

However, the human race is more varied still. Physicians consider an adult stature of between 1370 and 2010 mm to be 'clinically normal' (the exact figures quoted differ—this is of no consequence whatever since they are quite arbitrary). The tallest living person is Muhammad Alam Chamna of Analandar in Pakistan—he stands some 2510 mm. The shortest living adult is Antonio Ferreira, a Portuguese professional drummer, who is 750 mm tall (McWhirter 1984).

3.1. *Sex differences*

Human beings come in two basic variations—the male and the female. A biologist would refer to this difference as sexual dimorphism. There is a tendency at present, particularly among social scientists, to replace the word 'sex' with 'gender'. We shall follow the precedent of Archer and Lloyd (1982) in using the word 'sex' to describe those differences between men and women which are predominantly biological in origin, and 'gender' to describe differences predominantly determined by social and cultural factors. Hence, to take two extreme examples, the possession of male or female reproductive organs is a sex difference whereas the wearing of male or female attire or bodily ornament is a gender difference. Anthropometric differences between men and women are much closer to the former extreme than the latter. We can be fairly certain that men and women differ in stature for biological (i.e., genetic) reasons. But what of physical strength? Is this determined by biological factors or social ones? The answer, as we shall see subsequently, has to be 'a mixture of both'.

What are the most useful statistical indices of sex differences? The most frequent approach throughout the literature is a straightforward comparison of means. Hence, we read statements like "on average women are 7% shorter than men" or "on average women are 65% as strong as men". Let us call the average female dimension (or strength) divided by the average male dimension (or strength) the F/M ratio for short. However, for all the variables we are likely to consider in the present text, there is considerable overlap between the male and female distributions. The F/M ratio of means tells us very little about this combined distribution. (Among many other equally interesting descriptions we might include the ratio of the 95th %ile female to the 5th %ile male or the 5th %ile female to the 95th %ile male; the percentage of women stronger than the 5th %ile man or men weaker than the 95th %ile woman, etc.).

At the very least, a descriptive index should reflect both the difference between the means and the magnitude of the variances of the male and female distributions under consideration. It would be useful and informative to know the proportion in the total variance in strength (i.e., in the combined unisex distribution) which is attributable to strength. Afficionados of the one-way analysis of variances will understand that this index is given by the equation:

$$R^2 = \text{between sex Ssq/total Ssq} \tag{3.1}$$

(If this equation is absolute gibberish to you don't worry too much—alternatively turn to any textbook of statistics.)

When preparing a paper on sex differences in strength a few years ago (Pheasant 1983), I paused to ask myself just what does the layperson have in mind when he (or she) asks "How true is it that men are stronger than women?" Consider a population of men and a population of women. Suppose we select a man at random followed by a woman at random and compare their strengths. We will call such a comparison a chance encounter. If we perform an infinite number of such comparisons we may generate a statistical distribution of chance encounters. The F/M ratio is equivalent to an encounter between an average woman and an average man. Both the layperson and the human scientist wish to know about the remainder of the distribution. For reasons which would only be comprehensible to a competent mathematical statistician, the distribution of the ratios of two normal distributions is not itself normally distributed. If, however, we forget about ratios and consider absolute differences the problem becomes much more tractable. If differences are used the distribution of chance encounters is normal and its parameters are given by:

$$M_e = M_m - M_f \quad (3.2)$$

$$S_e^2 = S_m^2 + S_f^2 \quad (3.3)$$

where the subscripts m and f refer to men and women, respectively. The value of zero in this distribution represents a chance encounter between a man and a woman of equal strength. It is simple to calculate the proportions of the distribution lying on either side of this point (by calculating z and looking up a table as described in Section 2.1). We therefore know the percentage of chance encounters in which the female is stronger or, in the most general sense, the percentage of chance encounters in which the female exceeds the male (% CEFEM). This index is as close as we can reasonably get to the layperson's conception of the question.

Obviously enough, any analysis of sex differences will founder if the male and female subjects are not truly comparable. Hence, a comparison of male navvies with female secretaries or male academics with female athletes is not a legitimate study of sex differnces.

Table 3.1 shows the values of these indices for a number of common anthropometric variables. These data, based on Stoudt *et al.* (1965, 1970), refer to the adult civilian population of the USA. In general, men exceed women in all the linear bodily dimensions except hip breadth. Women, however, exceed men in skinfold thickness.

There are ethnic differences in the magnitude of sexual dimorphism—at least for stature. Eveleth (1975) found greater differences in American Indians than in Europeans, who, in turn, showed greater differences than Africans.

Many sex differences in bodily proportions are too well known to require any

Table 3.1. Sex differences in bodily size and shape.

	Men			Women					
Dimension	Mean	SD	CV (%)	Mean	SD	CV (%)	F/M (%)	R^2 (%)	%CE FEM
1. Stature (mm)	1732	69	4·0	1600	64	4·0	92	50	8
2. Body weight (kg)	76·3	12·6	16·5	64·5	12·6	19·5	84	18	25
3. Sitting height (mm)	904	38	4·2	846	37	4·4	93	37	13
4. Knee height (mm)	541	31	5·7	498	27	5·4	92	36	15
5. Buttock–knee length (mm)	592	31	5·2	566	31	5·5	96	15	28
6. Hip breadth (mm)	356	29	8·1	366	37	10·1	103	62	58
7. Shoulder breadth (mm) (biacromial)	396	20	5·1	353	20	5·7	89	55	6
8. Triceps skinfold (mm)	13	6·8	52·3	22	8·1	36·8	169	26	80
9. Subscapular skinfold (mm)	15	7·4	49·3	18	9·1	50·1	120	3	60
10. Sum of skinfolds (mm)	28	13·3	47·5	40	16·6	41·5	143	14	71
11. Sitting height index (%)	51·8	1·5	2·9	52·4	1·5	2·9	101	4	61
12. Biacromial breadth index (%)	22·5	1·3	5·8	21·7	1·2	5·5	96	9	33

comment whatever. In general, the lengths of the upper and lower limbs and their component parts proportionally, as well as absolutely, are greater in men. This is reflected in the distribution of the sitting height index given in Table 3.1. (The reader will recall that we have decided to call the ratio between the sitting height and stature in an individual the sitting height index and the ratio of the average sitting height of a sample of individuals to their average stature the relative sitting height—see Section 2.5.)

The only limb dimension which is proportionally greater in women is buttock–knee length. This, presumably, is due to the substantial soft tissue component of this dimension and the 'better upholstered' female buttock. It is noteworthy that there is no sex difference in the proportional values for either head length or head breadth.

In addition to the dimensional anthropometrics described above, men and women differ in their bodily composition. In general, fat represents a greater proportion of body weight in the adult female than in the male. (Subcutaneous fat is also distributed differently, women having a propensity to accumulate fat in the breasts, hips, thighs and upper arms. Abdominal fat accumulates above the umbilicus in men and below the umbilicus in women.) The most direct way of measuring body fat is by densitometry. Fat is a good deal less dense than lean tissue, so if the density of the body is determined (usually by underwater weighing) it is possible to calculate the percentage which fat contributes to the weight of the body. Durnin and Rahaman (1967) found this percentage to be 13·5 [5·8] for adult men and 24·2 [6·5] for adult women (F/M = 179%, R^2 = 43%; % CEFEM = 89).

I have previously published a detailed analysis of sex differences in strength (Pheasant 1983). A survey of the literature located a total of 112 datasets in which a direct and, presumably, valid comparison of the performances of men and women in some test of static strength could be compared. Indices of sex differences were calculated for each of these datasets (see Table 3.2). Although the average value of the F/M ratio is 61%—very close to commonly quoted figures of women being two-thirds as strong as men—the ratios found in the whole series range from 37 to 90%. The other indices tell a similar story—sex can account for a major (85%) or a

Table 3.2. Sex differences in strength (classified by part of body, etc., tested).

Part of body	No. of tests	F/M (%)			R^2 (%)			% CEFEM		
		Min.	Ave.	Max.	Min.	Ave.	Max.	Min.	Ave.	Max.
Lower limb	17	50	66	81	11	43	62	4	12	28
Push/pull/lift	41	38	65	90	3	33	72	1	14	37
Trunk	11	37	62	68	35	54	61	1	6	13
Upper limb	29	44	58	86	7	53	77	1	8	35
Miscellaneous	14	43	53	61	35	63	85	0	5	16
All tests	112	37	61	90	3	45	85	0	10	37

Data from Pheasant (1983).

negligible (3%) proportion of the total variance in strength.

An interesting pattern emerges if we divide the datasets into groups according to the part of the body tested. Upper limb tests show greater sex differences than lower limb tests, or tests of pushing, pulling and lifting actions, with tests of trunk strength being somewhere between the two. Subdivision of the upper limb category revealed that tests of hand and forearm muscles gave lesser sex differences than tests of upper arm and shoulder muscles. Taking all three indices together, the least sex differences were found in the push/pull/lift category. (The factors which determine performance in a task of this kind are numerous and complex. The co-ordinated activity of many muscles may be involved and in some cases the limiting factor may be body weight and its leverage.) Hettinger (1961), considering the magnitude of the F/M ratio in various muscle groups, suggested that the sex difference is small in those muscle groups which are underused in everyday life. In the present author's view it would be just as reasonable to propose the opposite hypothesis on the basis of the available data—and the opposite would be just as biologically plausible! What is the underlying physiology of sex differences in strength? The strength of a muscle is directly proportional to the effective cross-sectional area of its contractile tissue. The cross-sectional area of a muscle must be closely related to the bulk which is visible to the casual observer, or can be measured with a tape. Ikai and Fukanaga (1968), using a sophisticated ultrasound measurement of cross-sectional area, found strength of approximately 6·5 kgf/cm^2 of muscle tissue which was independent both of sex and of age from 12 years upwards. Trained judomen had the same strength per unit area as untrained people. In general then, it is the quantity not the quality of muscle that counts; at least in strength measurements of short duration.

It is widely accepted that the 'secondary sex differences' of fat distribution and muscle bulk result from the relative concentrations of the sex hormones—androgens in the male and oestrogens and progestogens in the female. Testosterone, the most important of the androgens, is produced in large quantities in the testis, but also in very small quantities in the ovaries. (The level of testosterone in the blood plasma of men is 20–30 times that of women.) In response to a given training programme, men show a faster and greater increase in strength than women; and this difference is generally attributed to the effects of testosterone (Hettinger 1961, Klafs and Lyon 1978). Brown and Wilmore (1974) monitored a small group of female throwing event athletes over a 6 months' maximal resistance training programme. Strength increased by 15–53% but there was little evidence of muscular hypertrophy (increase in bulk). How this latter finding is to be reconciled with the results of Ikai and Fukanaga (1968) is not yet clear, but it is worth recording that most dance teachers believe that strength may be achieved without bulk if the exercise programme emphasises stretching actions—this belief, however, has not yet been the subject of formal scientific investigation.

Klafs and Lyon (1978) speculate that women who have a relatively high level of plasma testosterone will 'bulk up' like men in response to intense weight training and, indeed, the current vogue for female body building shows that a striking degree of muscular hypertrophy may occur in some individuals. It has often been suggested that female athletes are in one respect or another less 'feminine' than their more sedentary sisters. Malina and Zavaleta (1976) calculated an 'androgyny score' of 3 biacromial (shoulder)–bicristal (pelvic) breadth. (Hence a high score on this index is indicative of a masculine skeletal frame and a low score is feminine.) Runners (both long and short distance) did not differ from non-athletes in androgyny scores but jumpers and throwers were significantly 'masculine' according to this criterion. Is this a training effect or does it represent self-selection? The latter is generally believed to be the case amongst physical educationalists (Klafs and Lyon 1978). Adams (1961) compared young black women who had been engaged in heavy farm labour all their lives with ones who had not. Although the labourers were larger in overall size and muscular development than the controls, the 3 biacromial–bicristal androgyny score was similar for the two groups. An aspect of the question which occurs to many people, but concerning which there is little scientific evidence, is the psychological one. Does the average girl or woman 'try' as hard in a strength test as the average boy or man? We do not know the answer to this—or, indeed, even how to investigate the question. The emergent pattern of this jigsaw of data suggests that the overlapping strength distributions of men and women are to some extent the consequences of hormonal predispositions for the acquisition of muscular tissue—but that differences in patterns of habitual activity may accentuate or diminish these differences. The acquisition of new pieces for the jigsaw will make the picture clear.

The ergonomic consequences of sex differences are so widespread that no special enumeration is appropriate here; they will be discussed at length in several subsequent chapters. Our discussion would not be quite complete without some passing reference to 'norms', 'ideals', 'cultural expectations' and the elusive phenomena of taste and preference. The history of European art reveals considerable diversity in the ideal female form—consider, for example, the way that Venus was depicted by Rubens, Titian, Botticelli and Cranach, to name but four in order of decreasing radius of curvature. The ideal male form (Mars, Adam, etc.) has remained remarkably constant by comparison (or is it just that I don't notice the differences?). The one empirical study of these matters that I have discovered is that of Garner *et al.* (1980) who took the strikingly original approach of analysing the recorded heights, weights and bodily circumferences of all *Playboy* magazine centrefolds between 1959 and 1978. The trend over the period was for an increase in height, reduction in weight for height, bust circumference and hip circumference, and increase in waist circumference—indicating a move towards a body form which the authors characterized as 'tubular'.

3.2. Ethnic differences

The ethnic diversity of mankind has fascinated anthropometrists since the early days of their subject. Until the relatively recent emergence of ergonomics, anthropometric data were principally collected in the interests of human taxonomy—that is, the classification of human beings by a consideration of the affinities and differences between samples drawn from the different 'races' of mankind. Over the years physical anthropologists have diligently measured human beings and their skeletal remains in all corners of the globe, collecting a bewildering variety of data which are scattered throughout a century or more of learned journals in several languages. I have searched in vain for a definitive compilation or catalogue of the large number of isolated datasets which may be presumed to exist. The reader who wishes to inquire into the subject of ethnic differences could do much worse than to commence with the classic introduction of Ashley-Montagu (1960).

An ethnic group may be defined as a sample or population of individuals who inhabit a specified geographical distribution and who have certain physical characteristics in common which serve, in statistical terms, to distinguish them from other such groups of people. These characteristics may be presumed to be predominantly hereditary, although the extent to which this is the case is often contentious.

Ethnic groups may or may not be co-extensive with national, linguistic or other boundaries—hence the various ethnic types to be found within the population of Europe are distributed across national boundaries, although the frequency with which a given type is encountered will vary from place to place. To some extent ethnic groups fall into more or less natural clusters, which may be referred to as the Negroid, Caucasoid and Mongoloid 'divisions' or 'major groups' of mankind. The term 'race' has tended to disappear from the scientific literature, due, one might suppose, to a collective embarrassment occasioned by its misuse for dogmatic and propagandist purposes. In fact, as Gould (1984) has clearly shown, supposedly objective scientific writers have colluded in this misuse.

The Negroid division includes most of the dark-skinned peoples of Africa, together with certain minor ethnic groups of Asia and the Pacific islands. The Caucasoid division includes both light- and dark-skinned peoples resident in Europe, North Africa, Asia Minor, the Middle East India and Polynesia (together with the indigenous population of Australia and some other ethnic groups who form a subdivision of their own). The Mongoloid division comprises a large number of ethnic groups distributed across central, eastern and south-eastern Asia, together with the indigenous populations of the Americas.

The continuous process of migration and intermarriage which has continued throughout history, and is likely to accelerate rather than diminish, ensures a state of flux which renders any taxonomic scheme at least partially arbitrary. Fortunately, the formal subdivision of the human race is of little relevance to our present purposes. Our concern will be more with 'samples' and 'populations' defined by various criteria of operational relevance and these will be named, as far as is reasonably possible, in plain language—with minimal recourse to assumptions concerning underlying ethnicity.

Samples drawn from the civilian or military populations of the USA (of which there are a considerable number in the literature) will be classed as 'of predominantly European descent'—notwithstanding the fact that around 10% of the membership of such samples are of identifiably different ethnic origins and that the genetic admixture of the remainder is rich and complex.

Samples of adults may vary from each other either in overall size (as measured by stature or weight) or in bodily proportions. The most characteristic ethnic differences are of the latter kind since the major divisions of mankind include both tall and short populations. Figure 3.2 illustrates some salient features. Average sitting height (measured from the seat surface) has been plotted against average stature. The ratio of the two (relative sitting height) is plotted as oblique lines on the chart. When relative sitting height is large, the sample is 'short legged' and vice versa. The data points are all male samples taken from Eveleth and Tanner (1976) and NASA (1978).

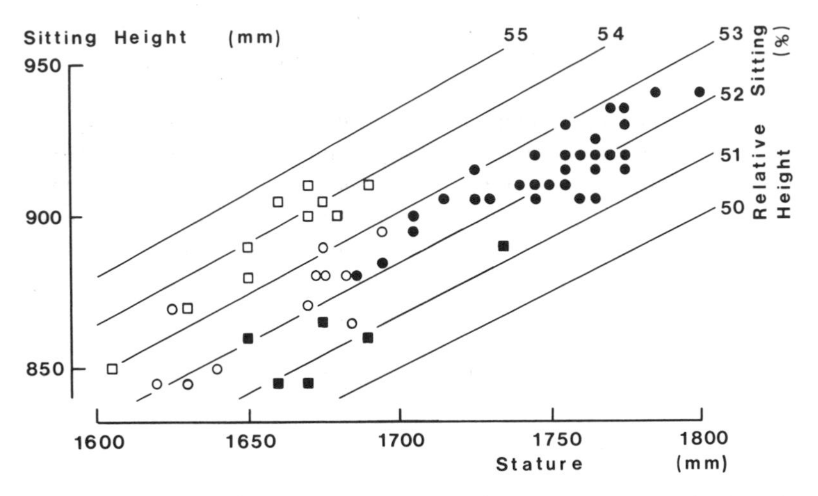

Figure 3.2. Ethnic differences in the relationship between average sitting height and average stature in samples of adult men.

● = European (including samples of predominately European descent); ○ = Indo-Mediterranean; □ = Far Eastern; ■ = African.

Black Africans have proportionally longer lower limbs than Europeans; Far Eastern samples have proportionally shorter lower limbs, the difference being most marked in the Japanese, less in the Chinese and Koreans and least in the Thai and Vietnamese. These differences of proportion occur throughout the stature range. If we consider the European data only, there is a tendency for the ratio of sitting height to stature to be slightly greater for short samples than tall ones—suggesting the interesting hypothesis that the lower limbs contribute more to differences in stature than the trunk. The populations of Turkey, the Middle East and India, labelled 'Indo-Mediterranean', have proportions similar to Europeans but, typically, a lesser overall stature.

Do these ethnic differences in size and proportion have any evolutionary significance? Zoologists have identified two rules concerning morphological variations of warm-blooded polytypic species—of which mankind is an example. Bergman's rule states that the body size of varieties increases with decreasing mean temperature of the habitat. Allen's rule states that the relative size of exposed portions of the body decreases with decreasing temperature. Roberts (1973), in an extensive survey of the anthropometric literature concerning the world's indigenous populations, showed that these rules are in general applicable to mankind. Body weight is negatively correlated with mean annual temperature. Samples with the lowest body weights are not found outside the tropics, and the highest body weights are not found at latitudes lower than 30°. Furthermore, linearity of bodily form (as indicated by high values for relative limb lengths) shows a strong positive correlation with mean annual temperature. Taken together these findings indicate that ethnic groups inhabiting hot climates will tend to have a high ratio of surface area to body mass—which is advantageous for the loss of heat. Similarly, the inhabitants of cold regions are adapted for heat retention. Roberts concluded, however, that there were differences in form between the major ethnic divisions of humanity even when the effects of temperature have been taken into account.

Although it is of little significance in ergonomics, no account of ethnic differences would be complete without mention of the cephalic index—the ratio of the maximum breadth of the head from side to side, to its maximum depth from front to back expressed as a percentage. A value of less than 75·9% is termed dolichocephalic, 76·0–80·9% mesocephalic, 81·0–85·4% brachycephalic and greater than 85·5% hyperbrachycephalic. Much has been made in the past of the cephalic index as a basis for the classification of ethnic groups. In general, African Negro populations are mainly dolichocephalic, but the other major divisions of humanity include groups which have a variety of head shapes. It should not be necessary to observe at this point that the shape (or, indeed, the size) of the head is not in any way indicative of the quality or quantity of the thought processes which occur within it—notwithstanding the contrary opinions which have been expressed in the past (Gould 1984).

The relative lengths of the upper limbs show a similar pattern of ethnic differences to the lower limbs, and there is some evidence to suggest that the differences are more due to a lengthening or shortening of the distal segment of the limb (i.e., the forearm or shank) than the proximal segment (i.e., the upper arm or thigh). The shoulders are a little narrower relative to stature in Africans than Europeans and the hips are considerably narrower in both sexes. In general, African bodily proportions are best described as 'linear'.

It would be a mistake to consider these differences in bodily size or shape to be fixed and immutable characteristics of ethnic groups. Several studies of migrant samples have shown significant differences between the growth patterns or adult dimensions of individuals born in the new environment and equivalent samples in the 'old country'. Boas (1912) and Shapiro (1939) are classic studies of this kind and subsequent investigations include Kaplan (1954), Greulich (1957) and, more recently, Koblianski and Arensburg (1977). Shapiro (1939) studied Japanese immigrants to Hawaii. He showed that although the Hawaiian-born generation are taller than the immigrants, and larger in most other dimensions, the ratios of the major bodily dimensions (i.e., relative sitting height, relative biacromial breadth) are not very different. This relative constancy of proportion has been confirmed by Miller (1961). This led Roberts (1975) to conclude that "the data suggest a strong genetic component to body proportion, and a more labile overall size". If Roberts's conclusion is correct, it is even more remarkable that several studies have shown differences between migrants and their descendants in the cephalic index—once considered a definitive indicator of ethnicity. Boas (1912) and various other authors have shown substantial shifts in the cephalic indices of the many ethnic groups which go to make up the population of the USA. These are sometimes increases, sometimes decreases and do not seem to converge on any single value (although it has been stated otherwise). The nature of the environmental factors which achieve these transformations within a generation remain utterly obscure.

The overall situation is yet more complex. Within one population, at least, secular trends in bodily form are occurring over and above the generally observed tendency towards increased body size (see Section 3.5). Yanagisawa and Kondo (1973) have demonstrated a marked secular increase in the cephalic index of the Japanese—from 83·0% in the 50–59-year-olds of their cross-sectional sample to 86·2% in the 18-year-olds, compared with an average index of 80·8% for Japanese students between the years 1910 and 1917. There is no evidence for such a secular trend in comparable European studies (e.g., Marquer and Chalma 1961). Furthermore, both Yanagisawa and Kondo (1973) and Tanner *et al.* (1982) have demonstrated a secular increase in the relative lengths of the lower limbs of the Japanese over the last few decades. Hence, while the bodily proportions of the Japanese are becoming more like Europeans, their head shape is diverging in the direction of more extreme brachycephalisation.

We must conclude that all ethnic differences are transient phenomena, subject to a variety of influences which we do not in the least comprehend.

How do these between-sample differences compare with the variation which may be found within ethnic groups? The question is not so easy to answer as it was in the case of sex differences (where the human race is divided into two categories rather than the many ethnic groups of mankind). Tildesley (1950) evaluated the anthropometric literature of her day. The results which Tildesley published allow an approximate calculation of R^2—in this case the proportion of the total variance in a characteristic which is statistically attributable to ethnicity. The results of some of these calculations—presented in Table 3.3—show that R^2 is typically close to one-half.

Before leaving the subject of ethnic differences, it is necessary to consider their practical consequences. How important, for example, is the presence of ethnic minorities in the target population for whom we are designing a certain piece of equipment? Supposing we are concerned with knee height in a certain design problem. Figure 3.3 shows what would happen when a target population originally consisting of British men ($m = 545$, $s = 31$) is modified by increasing numbers of a short-legged group (Japanese, $m = 495$, $s = 21$) or a long-legged

Table 3.3. Approximate proportions (R^2) of overall variance in anthropometric data attributable to ethnicity.

Dimension	Overall[a] Mean (mm)	SD (mm)	R^2 (%)
Stature	1652	77	42
Sitting height	864	46	48
Span	1740	94	41
Shoulder breadth (biacromial)	370	28	50
Pelvic breadth (bicristal)	280	20	34
Head length	188	8	38
Head length	150	8	53

[a]The overall mean and SD are the parameters for all surveys of adult men lumped together—they are the best ultimate available of the distribution of these dimensions in all adult male humanity. R^2 is a statistical index, as defined in Equation 3.1, of the percentage in overall variance which is attributable to differences between samples, that is, to ethnicity.

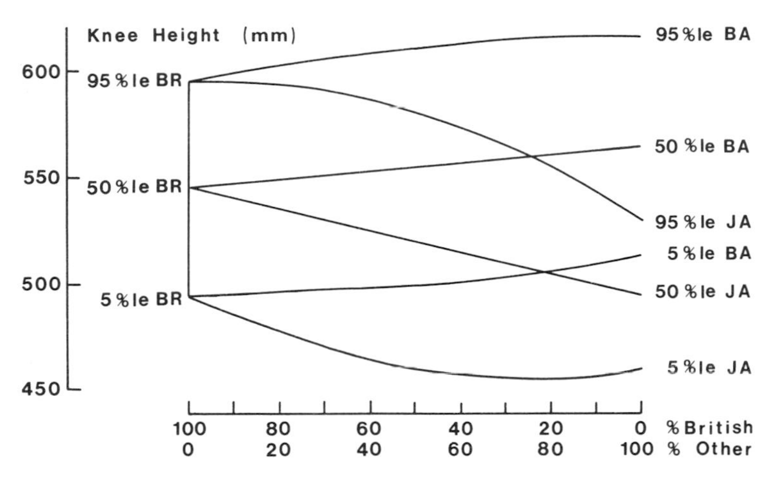

Figure 3.3. Effects, on the percentile values of knee height, of varying the relative numbers of British (BR), Japanese (JA) and black American (BA) men in a target population of mixed ethnicity.

group (black Americans, $m = 565$, $s = 32$). Figures were calculated using Equations 2·25 and 2·26. (The minimum value through which the 5th %ile Japanese curve appears to pass is an artefact due to the continued assumption of normality in a situation in which the true distribution is platykurtic—see Section 2.1.) The actual figures are only approximate but they suggest that in the case of the Japanese the differences that might interest a furniture designer (25 mm) occur at the level of around 30% membership, whereas for the black Americans the differences are of marginal consequence. There may, of course, be situations in which ethnic differences are much more critical. The research of Thompson and

Booth (1982) suggests that there are circumstances in which workers from certain ethnic minorities might be more at risk if industrial safety guarding standards are not modified to take into account their particular anthopometric characteristics.

3.3. Growth and development

At birth we weigh some 3·3 [0·4] kg, and we are 500 [20] mm in length, of which our trunks represent some 70%. In the two decades which follow our body length increases between three- and fourfold, our weight increases around 20-fold and our linear proportions change so that in the adult state the length of the trunk accounts for only 52% of the stature. However, the adult condition is by no means stationary—our bodily proportions are modified by our lifestyles and the inevitable processes of ageing. The anthropometrist who wishes to chart this course (or part of it) may most conveniently do so by a cross-sectional study in which several samples of individuals, representative of different age bands, are measured at the same time. (A cross-sectional age-band sample is known as a 'cohort'.) Data gathered by this means have certain limitations. In the case of children, only a very crude estimate can be obtained of the rate at which changes are taking place. Furthermore, our differences may be confounded with the effects of a secular trend. To disentangle these effects it is necessary to conduct longitudinal studies in which a sample of people are followed over an extended period of time. In practice, a mixed cross-sectional and longitudinal approach is often the most efficacious.

The genetic and environmental factors which control human growth have been documented in detail by Tanner (1962, 1978), who has also published standards for the height and weight of British children which have been widely adopted in medical practice (Tanner *et al.* 1966, Tanner and Whitehouse 1976). The pattern of growth of a 'typical' boy and girl based on these data is shown in Figure 3.4. (The 'typical' child is a purely fictitious individual who is average in all respects at all ages.) At ages up to 2 years measurements are made on a supine infant; subsequently in a standing position. The rate of growth in boys is very rapid during infancy, declining steadily to reach its minimum at 11½ years; it then accelerates again to reach its peak at 14 years before steadily decelerating as maturity is approached. The velocity peak around 14 years, known as the 'adolescent growth spurt', is associated with the events of puberty. The peak in the chart is broader and lower than it would be for any actual individual child since it represents the average of a sample of boys, all of whom are accelerating at different times. Hence, at 14 years some boys will have almost completed their growth spurt whereas others will scarcely have commenced it. As a consequence the standard deviations of the bodily dimensions of samples of adolescents are very large (see the tables in Chapter 4). The typical girl is a little shorter than the typical boy from birth to puberty but the growth spurt commences earlier in girls—at around 9 years reaching its maximum velocity at around 12 years; growth being more or less complete by 16 years. Hence, there is a period from about 11 to 13½ years when the

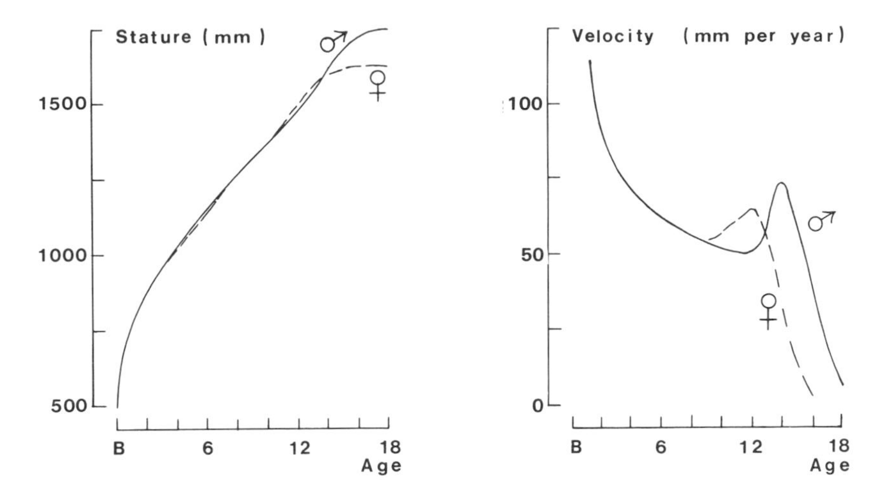

Figure 3.4. Growth from birth to maturity of a typical boy and girl: stature (left) and velocity (i.e., rate) of growth in stature (right).
Data from Tanner et al. *(1966).*

typical girl is taller than the typical boy. It might surprise many of us to learn that the typical boy reaches half his adult stature a few months after his second birthday, and the typical girl a few months before, although these figures will, of course, be subject to considerable variations in the population as a whole.

In addition to increasing in size, the human body changes considerably in shape. If the shape and composition of the body were the same throughout we would expect body weight to grow with the cube of stature (since weight is directly proportional to volume, assuming constant density). That would give an individual of average birth size, and who achieved an average male adult stature of 1740 mm, a body weight of 139 kg—which is close to twice the correct figure. In reality, growth is accompanied by an attenuation of bodily proportions.

Tanner (1962) has pointed out that there are various 'maturity gradients' which are superimposed upon the growth curve of the body as a whole—hence, at any point in time the upper parts of the body (particularly the head) are closer to their adult size than the lower parts; the upper limbs are further developed than the lower; and the distal segments of the limbs (hands, feet) are ahead of the proximal (thighs, arms). Cameron *et al.* (1982) also showed differences in the timing of the adolescent growth spurt for different parts of the body.

It is generally assumed that these gradients operate in such a way as to give a steady unidirectional transition from the large-headed, short-legged form of the child to the typical proportions of the adult. Tanner (1962) copied an illustration of this from Medawar (1944), who in turn took his from an anatomy textbook of 1915—which in turn is based on nineteenth-century data.

Medawar (1944) made the following statement: "Just as the size of the human being increases with age, so, in an analagous but as yet unformulated way, does his shape. The property is best expressed by saying that change of shape keeps a certain definite trend, direction or 'sense' in time; like size, it does not retrace its steps." Numerous authors have fitted mathematical equations to these supposedly simple transformations and some have attached biological significance to the constants in the equations.

Quite recently, whilst compiling anthropometric estimates for British schoolchildren (Department of Education and Science 1985), I chanced on certain discrepancies which lead me to wonder whether this simple scheme of things was justified. Figure 3.5, previously published in Pheasant (1984 a), is based upon the cross-sectional study of the under-18-year-old population of the USA published by Snyder *et al.* (1977). The mean value of each dimension for each age cohort has been divided by the mean value of stature (or supine crown–heel length for the under-2-year-olds). In some dimensions, such as head length, we may observe the smooth unidirectional approach towards adult proportions which we have been led

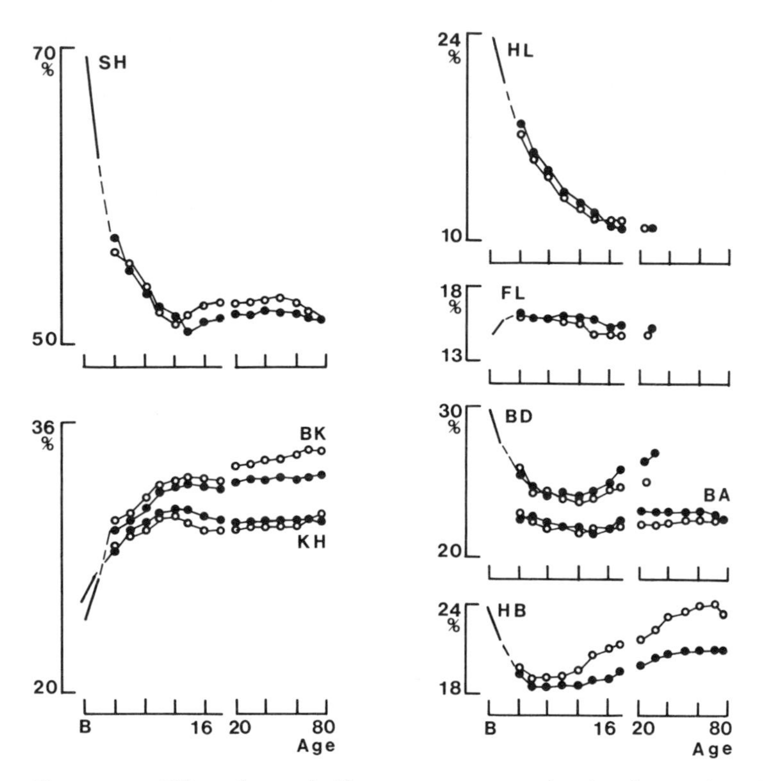

Figure 3.5. Effects of age on bodily proportions expressed as the relative values of various dimensions (% stature).
SH = sitting height; BK = buttock–knee length; KH = knee height; HL = head length; FL = foot length; BD = bideltoid breadth; BA = biacromial breadth; HB = hip breadth; ● = boys and men; ○ = girls and women.
Original data from Snyder et al. *(1977), Stoudt* et al. *(1965, 1970) and NASA (1978).*

to expect—but these are the exceptions rather than the rule. Most dimensions show what might be termed a 'developmental overshoot'. Sitting height, for example, has achieved its adult percentage of stature by 9 years in girls and 11 years in boys; it then overshoots and reaches a minimum at the time when the adolescent growth spurt is at its peak (12 years in girls, 14 years in boys), before climbing back to its adult proportions. Knee height, as one might expect, shows a pattern which is similar but inverted, as to a lesser degree do shoulder–elbow and elbow–fingertip lengths (not shown in the figure). Both shoulder and hip breadths are proportionally large in early infancy and pass through proportional minimum—during adolescence in the former case and childhood in the latter. Foot length has a long plateau of elevated proportions in childhood before commencing a descent during adolescence. In summary, the data confirm the popular stereotypes of the 'dumpy' infant and the 'gangling' adolescent.

The data of Figure 3.5 are also interesting with respect to sex differences and the ages at which the bodily proportions of boys and girls first diverge. In the case of sitting height, knee height and foot length the divergence is associated with the events of puberty and the developmental overshoot. The bony pelvis of the female is broader than that of the male at birth (Tanner 1978) and there is a slight sex difference in proportional hip breadth at the youngest age for which we have data—hip breadth also shows a slight divergence at around 6 years and a pronounced one at adolescence, which continues well into adulthood. (Buttock–knee length is quite similar so we are certainly dealing with soft-tissue upholstery to a large extent.) By contrast, shoulder breadth (bideltoid, biacromial) does not show any measurable divergence until as late as 17 years.

The muscular strengths of boys and girls are similar during childhood and diverge at around the time of puberty, as shown in Figure 3.6 which is based on the data of Montoye and Lamphier (1977).

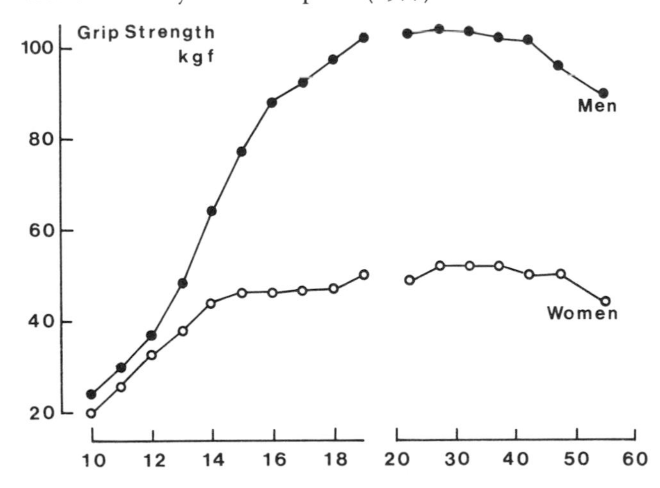

Figure 3.6. Effects of age and sex on grip strength. Data from Montoye and Lamphier (1977).

The age at which we reach 'anthropometric adulthood' is by no means as clearcut as one might suppose. Growth standards conventionally stop at 16 years for girls and 18 years for boys. The growth of a long bone occurs by cell division in plates of cartilage which separate the ends (epiphyses) from the shaft (diaphysis)—when this cartilage finally turns into bone, growth ceases (epiphyseal fusion). The clavicles continue to grow well into the twenties and so, to a lesser extent, do the bones of the spine. Andersson *et al.* (1965) demonstrated an increase in sitting height in a majority of boys after 18 years and girls after 17 years and in some boys after 20 years. A sample of Americans studied by Roche and Davila (1972) reached their adult stature at a median age of 21·2 years for boys and 17·3 years for girls; but some 10% of boys grew after 23·5 years and 10% of girls after 21·1 years. According to Roche and Davila (1972) this was partially due to late epiphyseal fusion in the lower limbs and partially to lengthening of the spine. Miall *et al.* (1967), in a longitudinal study of two Welsh communities, found evidence that men might grow slightly in stature well into their thirties.

3.4. The secular trend

Human biologists use the term 'secular trend' to describe alterations in the measurable characteristics of a population of human beings occurring over a period of time. Over a period of at least a century biosocial changes have been occurring in the population of much of the world which have led to:

increase in the rate of growth of children;
earlier onset of puberty—as indicated by menarche (the onset of the menstrual cycle) in girls and the adolescent growth spurt in both boys and girls;
increase in adult stature, with a possible decrease in the age at which adult stature is reached.

The extensive statistical evidence concerning these changes has been reviewed by, amongst others, Tanner (1962, 1978), Meredith (1976) and Roche (1979).

Tanner (1962, 1978) summarises the available evidence and concludes that from around 1880 to at least 1960, in virtually all European countries (including Sweden, Finland, Norway, France, Great Britain, Italy, Germany, Czechoslovakia, Poland, Hungary, the Soviet Union, Holland, Belgium, Switzerland and Austria), together with the USA, Canada and Australia, the magnitude of the trend has been similar. The rate of change has been approximately:

15 mm per decade in stature and 0·5 kg per decade in weight at 5–7 years of age;
25 mm and 2 kg per decade during the time of adolescence;
10 mm per decade in adult stature.

This has been accompanied by a downward trend of 0·3 years per decade in the age of menarche. Roche (1979) points out that secular changes in size at birth have been small or non-existent.

Although the magnitude of changes in Europe and North America have been fairly uniform they are by no means universal. Japan, for example, has shown a particularly dramatic secular trend. The data of Tanner *et al.* (1972) show that in the decade between 1957 and 1967 Japanese boys increased in stature by:

31 mm at 6 years;
62 mm at 14 years;
33 mm at 17 years.

In the 1967–1977 period, however, these figures had declined to:

17 mm at 6 years;
35 mm at 14 years;
19 mm at 17 years.

This suggests that the explosive biosocial forces driving the change are beginning to wear themselves out. In contrast, Roche (1979) cites evidence that in India, and elsewhere in the Third World, there has actually been a secular decrease in adult stature.

If people are increasing in size are they also changing in shape? The remarkable Japanese secular trend seems to be associated with an increase in the relative length of the leg—as the data of Tanner *et al.* (1982), plotted in Figure 3.7, show. It is doubtful, however, whether such a change of proportion is general. Figure 3.8 shows the relative sitting heights of samples of young American males (average ages between 18–30 years) plotted against the year in which the measurement was taken. There is no evidence of a secular trend in adult proportions. (This conclusion has been confirmed by Borkan *et al.* (1983).)

It is interesting to speculate as to whether our distant forebears were as short as we might imagine from recent secular trends. Anecdotal evidence concerning a range of artefacts from doorways to suits of armour abounds. Although it is not possible to calculate stature accurately from poorly preserved skeletal remains, the long bones of ancient burials allow us to make a reasonable estimate. The archaeological evidence summarised by Wells (1963) suggests that the statures of British males from neolithic to medieval times have always fallen within the taller part of the present-day human race. Indeed, figures quoted include average heights of 1732 mm for Anglo-Saxons and 1764 mm for Round Barrow burials; the latter actually exceeding the average height of present-day young men. The secular trend then seems to be a recovery from a setback which occurred somewhere in post-medieval times. Tanner (1978) cites various evidence that in the earlier part of the nineteenth century trends were small or absent, and plausibly associates them with the Industrial Revolution.

What then are the determining factors which have led to the phenomenon of the secular change? Speculation has been intense on this subject—most writers maintaining a cautious tone in their conclusions. Social/environmental influences such as the improved nutritional quality of diet and the reduction of infectious disease by improved hygiene and health care are the factors which most readily

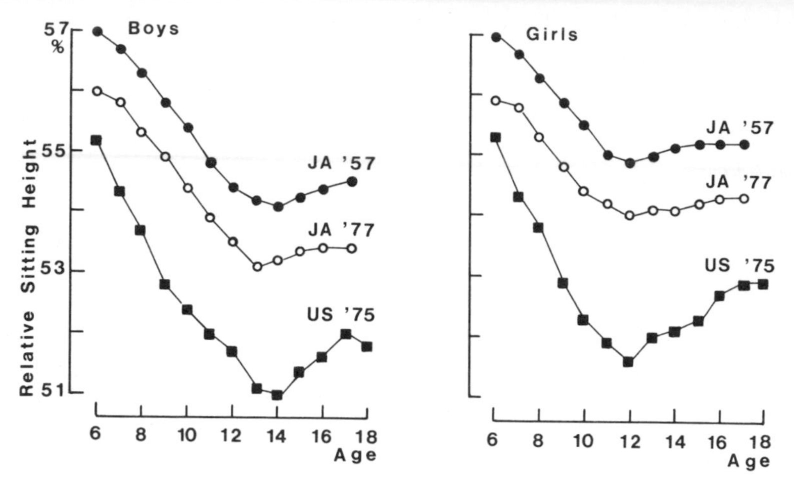

Figure 3.7. Secular trend in the bodily proportions of Japanese children (JA) compared with those of US children.
Original data from Tanner et al. *(1982) and Snyder* et al. *(1977).*

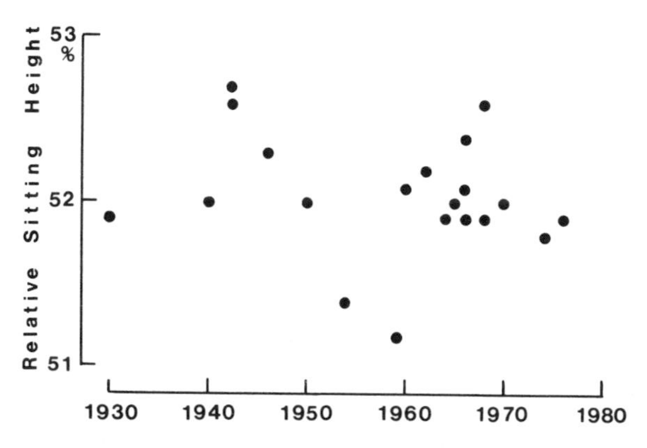

Figure 3.8. Relative sitting heights of samples of young adult US men measured between the years 1930 and 1980. Note the absence of any detectable secular trend in bodily proportions.

spring to mind, to which we might add the effects of urbanisation and reduced family size, but we cannot ignore the possible influences of genetic factors such as heterosis—the beneficial effects which are said to derive from outbreeding and the breaking up of genetic isolates. A century ago most people married and raised their children within the confines of isolated communities—today we are approaching the condition of the 'global village'. As Tanner (1962) perspicaciously observed "it has been shown in several West European countries that outbreeding has in fact increased at a fairly steady rate since the introduction of the bicycle".

The consensus view amongst human biologists tends to favour the environmental rather than the genetic causes. It seems most likely that genetic endowment sets a ceiling level to an individual's potential for growth and that environmental circumstances determine whether this ceiling is actually reached. If this is indeed the case, the end of the secular trend is in sight, at least in the economically developed countries of Europe, North America and elsewhere—since we could reasonably argue that the further amelioration of environmental conditions, beyond those adequate for the achievement of full genetic potential, cannot lead to any further changes. Considerable evidence suggests that this limit has indeed been reached, at least in some communities. Backwin and McLaughlin (1964) showed that Harvard freshmen from relatively modest social backgrounds increased in stature by around 40 mm from 1930 to 1958, whereas those from wealthy backgrounds showed no change. Cameron (1979) has published data showing, very convincingly, that the secular trend for stature had levelled off, for children attending schools in the London area, by about 1960 (Figure 3.9). Tanner (1978)

Figure 3.9. Secular trend in the average stature of children in the London area. Note that changes have been minimal since 1960.
Data from Cameron (1979).

also showed that the secular decrease in the age of menarche had come to a halt by about this time both in London and in Oslo. The subsequent national survey of Rona and Altman (1977) confirm the impression that in Great Britain the secular trend has now reached a steady state. Rona (1981) was prepared to conclude that "there is no evidence that the secular trend in growth has continued after 1959 in the UK". Similarly, Roche (1979) reported that national surveys of US children and youths in 1962 and 1974 show constancy of stature (except at the 5th and 10th

%ile levels, where small increases have occurred). The absence of a secular trend admits of two possible interpretations. The optimistic is that conditions for growth have been optimised and that all children are now reaching their genetic ceilings. The pessimistic is that the percentage of children raised under optimal environmental conditions has simply ceased to increase. The continued existence of significant social class differences in growth (see Section 3.6) tends towards the pessimistic interpretation.

3.5. *Ageing*

Figure 3.10 shows the average heights and weights of the adult civilian populations of Great Britain and the USA plotted against age. A steady decline in

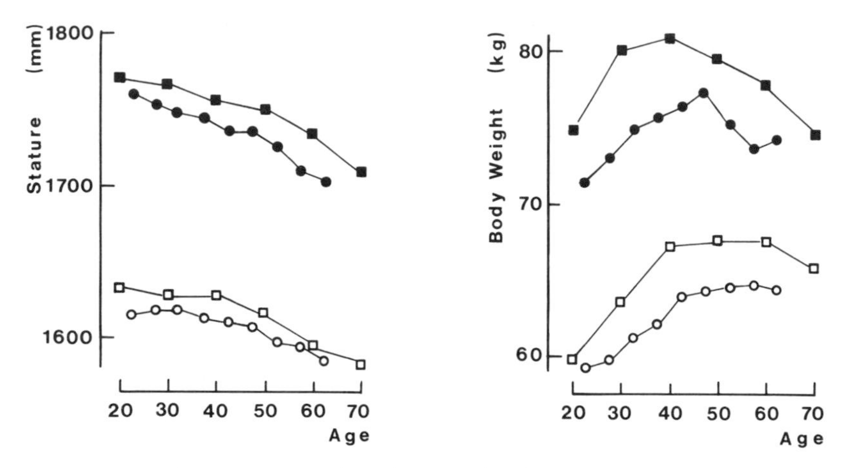

Figure 3.10. Average stature and weight in samples of adults of various ages.
■ = *men, USA;* ● = *men, Britain;* □ = *women, USA;* ○ = *women, Britain.*

stature is apparent, whereas weight climbs steadily before subsequently declining at around 50 years in men or 60 years in women. In analysing such a pattern we must consider the combined effects of the ageing process and the secular trend, together with the possibilities of differential mortality (i.e., that people with certain kinds of physique may tend to die younger). Damon (1973) showed that men of average height and weight had greater longevity than those who deviated strongly in either respect. These interactions require multicohort longitudinal studies for their elucidation. Investigations of this kind include the Welsh study of Miall *et al.* (1967) and the extensive Boston programme of the Veterans Administration (Damon *et al.* 1972, Friedlander *et al.* 1977, Borkan *et al.* 1983). Longitudinal studies show that at around 40 years of age we begin to shrink in stature, that the shrinkage accelerates with age, and that women shrink more than men. The shrinkage is generally believed to occur in the intervertebral discs of the spine—resulting in the characteristic round back of the elderly (e.g., Trotter and Gleser 1951)—although Borkan *et al.* (1983) suggest that some decrease also occurs in the lower limbs. The data show a longitudinal increase in weight for height until 55 years followed by a decline. Friedlander *et al.* (1977) showed a steady longitudinal increase not only in hip breadth but also in the bi-iliac breadth of the bony pelvis. The mechanism of the latter is obscure but it suggests that 'middle age spread' may not be totally due to the accumulation of fat, but may also involve changes in the bony pelvis.

If we assume that no secular change has occurred in the proportions of the body, then proportional ratios calculated from cross-sectional studies should be comparable with the longitudinal results. Figure 3.5 plotted from the data of Stoudt *et al.* (1965, 1970) shows that this is indeed the case. The proportional decrease in sitting height is compatible with the spinal shrinkage explanation of stature decline, and the greater change in women matches the longitudinal findings of Miall *et al.* (1967). Dimensions with a substantial soft tissue component such as hip breadth and buttock–knee length show a proportional increase (until 75 years) which is more pronounced in women. The proportional decline in the biacromial breadth of men over 60 years old presumably reflects the characteristic rounding of the shoulders of the elderly. It is of interest that for both sitting height and biacromial breadth the ageing process finally abolishes the sex difference altogether.

Cross-sectional studies such as Stoudt *et al.* (1970) have shown an increase in skinfold thickness followed by a decline at around 40 years in men and 60 years in women. There is evidence, however, that this represents a redistribution rather than a loss of body fat. Durnin and Womersley (1974) showed that the relationship between skinfold thickness and whole body fat, as measured by densitometry, changes with age. It seems there is a transfer of fat from subcutaneous positions to deep ones (e.g., around the abdominal organs). The net quantity as a percentage of body weight continues to increase, and the longitudinal decline in weight that we see late in life is probably due, therefore, to the loss of lean tissue. Borkan and Norris (1977) found that the weight of fat was constant with age in a cross-sectional sample of middle-aged men, but that lean tissue declined markedly. Subcutaneous fat decreased on the trunk but increased on the hips, but this was accompanied by an increase of abdominal (waist) circumference indicative of a sagging of the abdominal contents (due presumably to increased internal fat and decreased muscular resistance). A similar redistribution presumably occurs in women—but there is little numerical evidence.

The above studies are all based on the populations of the USA and Great Britain, where obesity, consequent upon an abundant food supply and a sedentary lifestyle, is prevalent. The situation in other communities will, of course, be different, and in societies where food is scarce, adult increase in body weight does not occur.

The loss of lean body weight is due principally to a wasting away of muscles (although the bones also become less dense in later life) and this leads to a decrease in muscular strength. Numerous cross-sectional studies have shown that strength continues to increase in early adulthood but declines in middle age and thereafter (Hettinger 1961). The data of Montoye and Lamphier (1977) plotted in Figure 3.6 are a fairly typical example of the effects of age on muscle strength. It is worth noting also that the mechanical tensile breaking strengths of bone, muscle and connective tissue also decline with age (Yamada 1970), possibly more rapidly than does voluntary muscular strength, hence increasing risk of injury. The flexibility of the joints of the body decreases dramatically with age, although I know of remarkably little empirical data concerning this rather obvious phenomenon. Furthermore, the incidence of disabling musculoskeletal disorders increases (see Chapter 10).

Trends in the birth and death rates enable government statisticians to anticipate future trends in the age structure of the population, and it is true to say that we live in an ageing society. The number of people in Great Britain aged 65 years or over grew by over 2 million between 1961 and 1984; the trend is, however, flattening out due to relatively low birth rates in the 1920s and 1930s. The over-75 age group will, however, increase substantially in numbers by the end of the century (Ramaprakash 1984). The quality of their lives will depend in no small measure on the equipment and environments which we design for them.

3.6. Social class and occupation

Social class and occupation are inextricably linked—so much so that the latter is generally used as an operational measure of the former. The widely used system of the Office of Population Censuses and Surveys, known as the 'Registrar-General's Classification', divides occupations into six categories: I Professional, II Intermediate, IIIA Skilled Non-manual, IIIB Skilled Manual, IV Semi-skilled Manual and V Unskilled Manual. Every so often it is necessary to re-classify an occupation as its perceived status changes.

In a fascinating study of primiparae (women pregnant for the first time) in Aberdeen, Thomson (1959) found that stature was stratified by the occupation of the subject's father, by her own occupation and by her husband's occupation but, most remarkably, that tall girls had a stronger tendency to marry upwards, with respect to their father's and their own occupations, than did short ones.

Social class differences in stature remain marked. Knight (1984), in a nationwide study of the adult population of Great Britain, found an average stature of 1755 mm for men and 1625 mm for women in social classes I and II as against 1723 and 1596 mm in social classes IV and V. The differences were of similar magnitude for all age cohorts. The pattern was less clear for body weight. The same survey also showed regional differences; ranging from an average stature of 1751 mm for men and 1619 mm for women in south-west England to 1719 and 1594 mm in Wales.

Extensive British data show class differences in the growth of schoolchildren.

Rona (1981) has reviewed the evidence of British surveys over the last 30 years or so. A difference of between 10 and 20 mm in average stature between the top and bottom classes in the Registrar-General's table already exists at 2 years. By 7 years this has widened to 30–40 mm, a gap that has remained constant for the last 30 years. In the most recent of these surveys, Goldstein (1971) and Rona *et al.* (1978), the differences between classes I to IV were relatively modest, but it seems that at present differences in primary-school children are mainly due to those in social class V. Rona *et al.* (1978) showed that the children of unemployed fathers are shorter in stature, but that the parents of these children were also shorter within each social class. Lindgren (1976) reported an extensive survey of urban children in Sweden between 10 and 18 years of age. There was no difference in height, at any age, between the social classes as defined either by the father's occupation or family income. Sweden is the only country in the world where this is known to be the case—a fact which Tanner (1978) takes to be an operational and biological measure of the existence of a 'classless society'.

Let us now turn to the anthropometric characteristics of specific occupational groups and of samples defined similarly by other activities such as 'students' or drivers. Some occupations (policemen, policewomen, aircrew) impose explicit height limits as part of their selection criteria. In other cases the selection may be implicit in the social strata from which the sample is drawn. Table 3.4 shows some examples of the latter. Comparing the British population of adults in general (Knight 1984) with subgroups of approximately the same age—drivers, medical students and BP employees—an interesting sex difference emerges. The between-sample differences are greater for women than they are for men, suggesting that a greater selective pressure on entry to these 'occupational' groups exists for women than for men. It also urges caution in the selection of anthropometric data 'representative' of the population.

Table 3.4. Statute of British adults—differences associated with age, sex and occupation.[a]

		Men				Women			
Occupation	Age	5th %ile	50th %ile	95th %ile	SD	5th %ile	50th %ile	95th %ile	SD
General population	19–65	1623	1738	1873	70	1506	1608	1710	62
	19–25	1640	1760	1880	73	1518	1618	1718	61
Motor-car drivers	17–65+	1626	1738	1850	68	1533	1625	1717	56
	17–24	1649	1753	1857	63	1550	1641	1725	55
BP employees	20–69	1641	1751	1861	67	1516	1623	1730	65
	20–24	1655	1766	1876	67	1540	1636	1731	58
Medical students	18–24	1668	1770	1871	62	1546	1650	1754	63

[a]In the younger age group the proportion of variance (R^2) in stature attributable to occupation is 0·3% for men and 3% for women.

Data from Knight (1984), Haslegrave (1979), Haslegrave and Hardy (1979), Pheasant (1984 b) and Montegriffo (1968).

In some circumstances 'self-selection' may occur, with individuals gravitating towards professions to which their physique is well suited. The physical content of the occupation itself may also exert a training effect—or, perhaps more importantly, a de-training effect in the case of sedentary lifestyles. (The most extreme examples of these effects are provided by athletes—see Wilmore (1976).)

A classic study of 'selection and de-training' is that of Morris *et al.* (1956), who investigated the waist and chest girths of the uniforms of London busmen—both drivers and conductors—ranging in age from 25 to 64 years. In addition to seeing a steady increase with age in both groups, the girths of the drivers were greater than those of the conductors—even in the youngest age group. The authors postulated therefore that "the men have brought these differences into the jobs with them". Passing from the ridiculous (if we may so term the busmen's trousers) to the almost indisputably sublime, two studies of ballerinas are worth mention. Grahame and Jenkins (1972) measured the joint flexibility of a group of female ballet students and found it to be greater than controls, even for joints, such as those of the little finger, which were not trained as such. The authors therefore concluded that only girls gifted with generalized joint flexibility would undertake the rigours of ballet training. Vincent (1979) has documented the appalling, and sometimes disastrous, lengths to which these girls will sometimes go in 'competing with the sylph'—that is in pursuit of the unnaturally slender body form which is sufficiently otherworldly for their art.

Chapter 4
Static anthropometric data

The data of the present chapter are all 'static': the dimensions are linear lengths, breadths or depths of the human body. The tables represent the culmination of a lengthy programme of research, compilation and editing, the principal objective of which has been to provide as complete a picture as possible of human diversity in size and shape as applied to design problems.

4.1. *Accuracy*

At the outset we must clearly acknowledge the limitations of the dataset. Contemporary resources of anthropometric data do not permit the presentation of tables in which one has complete confidence with respect to accuracy—except for a certain limited number of dimensions or for restricted (usually military) populations. Accuracy is desirable but neither realistic nor entirely necessary for most practical design problems. (Most, but not all, of my professional colleagues would agree with the last statement.) In compiling a data manual such as this, one is faced with a straight choice. One can include only those data which are established, to a fine level of accuracy, beyond all reasonable doubt, accepting that by so doing only a very patchy picture will emerge, or one can fill in the very considerable gaps by techniques of extrapolation and estimation to paint a picture which, although somewhat impressionistic, is at least complete. (To pursue the analogy, one could compare several fragments of a landscape by Poussin or Constable with a completed Monet or Turner.) I have chosen the latter approach.

Why are we faced with such a dilemma? The answer lies in the difficulty and expense of conducting anthropometric surveys. Consider Tables 4.1–4.2 (see pp. 85–109) which describe the British population broken down into 26 age bands, the latter 21 of which are also broken down by sex, making a total of 47 'samples'. The accuracy of our data concerning such a sample would depend, *inter alia*, upon the number of subjects measured. Equation 2.17 tells us that if it is necessary to determine a given percentile within plus or minus u units, the number (n) of subjects required is given by:

$$n = ks^2/u$$

where s is the standard deviation of the dimension concerned and k is a constant for the particular percentiles we require. For the 5th and 95th %iles $k = 4 \cdot 14$. Supposing we wish to know the 5th and 95th %iles of our samples to an accuracy of ±5 mm. The standard deviation of the dimensions will, of course, be variable but experience might lead us to believe that a value of 40 mm would be typical. Hence, from the above equation, just over 1000 subjects are required; but we require this number in each of over 40 samples. Of course, we might say that an accuracy of ±5 mm is not actually necessary or an estimated standard deviation of 40 mm is overgenerous—our required resources would diminish accordingly. Supposing that by the most parsimonious design of our survey the number was reduced to a mere 10 000. I asked a fellow ergonomist who is experienced in such matters to estimate roughly what it would cost to plan and perform such a survey, analyse the data and prepare the tables over a 3–4 year period. The figure quoted was £300 000 at mid-1984 prices. Until such a survey is performed, those of us who do not design military equipment must make do with estimated data.

Most of the tables which follow were prepared by the method of ratio scaling described on p. 39. A detailed validation study of this technique is described in Pheasant (1982 a). The 1st and 99th %ile values of some 136 dimensions drawn from six different surveys were estimated from a knowledge of only the parameters of stature in the survey concerned. (This is a more rigorous test than the 5th and 95th %iles commonly used in design work.) The errors which accrued were random and conformed approximately to a normal distribution with parameters of −3 [13] mm. Ninety-three per cent of the errors fell within the range of ±25 mm. In many cases the estimates were within the confidence limits of the original survey (as calculated from Equation 2.17).

Since the time of the original validation study every effort has been made to improve the estimates by the incorporation of new reference sources and by a careful consideration of the relevance of the original sources; leading to the rejection of many which were of doubtful validity. It is to be assumed that these changes represent improvements rather than otherwise. Without engaging upon another validation study we cannot be sure—but the cyclical process of modification and re-validation could in theory go on for ever and one has to stop somewhere.

What accuracy is actually required in anthropometric data? This is a very difficult question which must be studied at several different levels. In the purely formalized statistical sense we may consider what percentiles a given percentile which is erroneously quoted actually represent, e.g., if a 95th %ile was in error, the figure quoted might in truth represent the 93rd or 98th %ile; with a consequence of mismatching a greater or lesser percentage of the target population in the design. In the validation study the estimates of the 1st and 99th %iles were checked in this way—on average the estimates would have included 96% of the population as against 98% for perfect data. It is more informative, however, to consider the likely errors of prediction alongside those which might arise in other ways. The human body has very few sharp edges—its contours are rounded and it is generally squashy and unstable. The consequent difficulty in identifying landmarks and controlling posture makes it virtually impossible to achieve an accuracy of better than 5 mm in most anthropometric measures—and for some dimensions the errors may be much worse (sitting elbow height is a notorious example). These errors, however, pale into insignificance in comparison with those which might occur in the application of even the most accurate tables. The questions of clothing corrections, postural variation and design criteria are discussed in subsequent sections.

At the end of the day we must weigh the benefits of having more accurate data against the costs of obtaining it. The data published here must be regarded as a set of provisional estimates. To paraphrase the words of a song "If this ain't data it will have to do—until the real thing comes along." Many ergonomists (myself included) would be more than willing to organize the definitive anthropometric survey of the British population (or any other one) for any sponsor who came up with the cash. Are our present doubts of any consequence? In the design of something like a protective face mask a high level of accuracy is required; for the majority of furniture, workstations or interior design problems it probably is not.

4.2. *Clothing corrections*

The great majority of anthropometric surveys are conducted on subjects who are nude or at most lightly clad. In contrast the artefacts we design are used by people who are clothed and shod in a variety of outfits—there are of course notable exceptions. Since the tables presented here are based on such surveys it will be necessary for the user to add clothing corrections appropriate to the specific application. The most important of these is an increment for the heels of shoes which should be added to all vertical dimensions which commence from the floor.

A minority of investigators, appalled by the arbitrary clothing corrections which have been advocated in some design guides, have decided to measure their subjects fully clothed and shod—arguing that for all practical purposes the outfit is part of the person (Andrew and Manoy 1972, Institute for Consumer Ergonomics 1983). Although this approach has certain *ad hoc* expediency, the data so obtained cannot be readily compared with other studies and its long-term value is hence diminished. Furthermore, clothed surveys are only relevant for target populations whose style of dress remains constant over a reasonably long period of time. In practice this means wearers of uniforms or groups who are socio-economically isolated from fashionable change—it would, for example, be worse than useless for female office workers. In many respects the best possible compromise is that adopted by Rebiffe *et al.* (1983), which is to make the crucial measurement of stature in both the clothed and unclothed conditions.

What clothing corrections should we apply to the data which follow? Two crude surveys of heel heights were carried out; the first in May 1981 and the second in May 1983. Data were obtained concerning the heel heights of those shoes which subjects wore most frequently on business or social occasions outside their homes (excluding 'specialist' items such as football boots, Wellingtons, etc.). The data may be modestly biased by an over-representation of young adults. The distributions of heel heights were as follows:

in 1981 men: 26 [11] mm; women: 55 [27] mm;
in 1983 men: 23 [10] mm; women: 40 [21] mm.

Women's heels had declined dramatically, due both to changes in the design of 'formal' shoes and to the increasing acceptability of sports shoes as fashion accessories. It is commonly believed that fashions in clothing and footwear follow more or less cyclic fluctuations. We hope that the furniture or interiors we design will continue in use during several such cycles; ideally, therefore, we should choose a correction which represents the midpoint about which the cycles oscillate. Assuming that the women's heels in the two surveys represent the peak and trough of such cycles, a figure of 48 [25] mm would be close to the midpoint. (Fashion conscious friends assure me that this assumption is a reasonable one.)

Theoretically, shoe corrections should be made according to Equations 2.39 and 2.40 (p. 38). Trying this for a few examples will convince us that corrections to the standard deviation are of negligible magnitude.

Taking one consideration with another, the following shoe corrections are recommended for most purposes:

add 25 mm to all percentiles for adult men, boys over 12 years old and all children under 12 years;
add 45 mm to all percentiles for adult women and girls over 12 years old.

The reader is free to use his discretion in modifying these figures wherever it may be appropriate.

Other clothing corrections are usually small—except for very heavy outdoor garments which may be relevant in some cases. Examples for guidance purposes are given in Section 4.4. Every so often a design application turns up in which additional accoutrements become central to the question—protective helmets, backpacks, etc., may be relevant in certain specialized areas. Much more widespread are the problems associated with the luggage carried by the passengers of public transport systems—as a consequence Williams (1977) performed what is believed to be the world's first 'luggometric' survey.

4.3. Standard anthropometric postures

The standardization of anthropometric measurement techniques has occasioned considerable discussion, much of which is of a very arid nature. The difficulty is principally historical, in that investigators working in different areas of application have evolved different technical traditions. The definitions of Hertzberg *et al.* (1963) have been particularly influential in ergonomics and could now claim to be standard. The present text follows them in most places. The reader is referred to Hertzberg (1968) for a succinct discussion of the standardization of anthropometric data.

The majority of the measurements in the tables which follow were made in one of two standard anthropometric postures (shown in the accompanying figures) which will now be described in detail.

The standard standing posture

The subject stands erect pulling himself up to his full height and looking straight ahead, with his shoulders relaxed and his arms hanging loosely by his sides. He stands free of walls, measuring instruments, etc. Some versions of this definition state that the 'Frankfurt Plane' of the head, which runs through the outer angle of the eye and the external auditory meatus, should be horizontal—this makes little difference in practice. (In some growth studies the investigator lifts the child's head in order to gently stretch him out; similarly in some adult studies the subject increases his stature by pressing himself against a wall. These 'stretch' techniques give different results from the standard ones and have not, in general, been adopted by ergonomists.)

The standard sitting posture

The subject sits erect on a horizontal surface, pulled up to his full height, looking straight ahead (or with his head orientated so that the Frankfurt Plane is horizontal). The shoulders are relaxed, with the upper arm hanging vertically and the forearm horizontal (i.e., the elbows are flexed at right angles). The height of the seat is adjusted (or supports are placed beneath the feet) until the thighs are horizontal and the lower legs vertical (i.e., the knees and ankles are flexed at right angles). Measurements are made perpendicular to two reference planes. The horizontal reference plane is that of the seat surface; the vertical reference plane is a real or imaginary plane which touches the back of the uncompressed buttocks and shoulder blades of the subject. The seat reference point (SRP) lies at the intersection of the vertical reference plane, the horizontal reference plane and the median plane of the body (i.e., the plane which divides it equally into its right and left halves).

One might reasonably object that these formal standardized postures are rarely used by real people in real activities. (As subsequent sections of this book progress we will see that this is not as great a problem as it first seems, since in many cases the data are used in conjunction with criteria defined by acceptable postural conditions.) To deal with this criticism several surveys have included measurements made in a 'normal slumped' sitting posture—in which one may suppose that the back is rounded and the shoulders are hunched to some variable extent. Indeed, I have included slumped data in previous compilations (Pheasant 1982 b, 1984 b). In retrospect, this seems needless. A direct subtraction of 40 mm from all percentiles of relevant adult dimensions should prove adequate to make what is, largely speaking, an arbitrary correction.

4.4. An annotated list of the dimensions in Tables 4.1–4.37

We shall now consider in turn each of the 36 dimensions which appear in the tables that follow. The notes concerning their applications will be brief but cross-referenced to other sections of this book in which they are discussed in greater detail. It may be assumed that all of these measurements refer to the standard postures described above.

The definitions are written both in the technical jargon of the anatomist-anthropologist and the language of the layperson. It is hoped that they will combine the precision of the former with the comprehensibility of the latter. The clothing and other corrections given are by no means definitive—they serve as examples only.

1. Stature

Definition: The vertical distance from the floor to the vertex (i.e., the crown of the head).

Applications: As a cross-referencing dimension for comparing populations and estimating data (see p. 39). Defines the vertical clearance required in the standing workspace; minimal acceptable height of overhead obstructions such as lintels, roofbeams, light fittings, etc.

Corrections: Shoes as in Section 4.2; 25 mm for a hat; 35 mm for a protective helmet.

A few design applications call for supine or prone body length (in which the subject lies on his back or front, respectively). Such a position lengthens the adult body by approximately 15 mm.

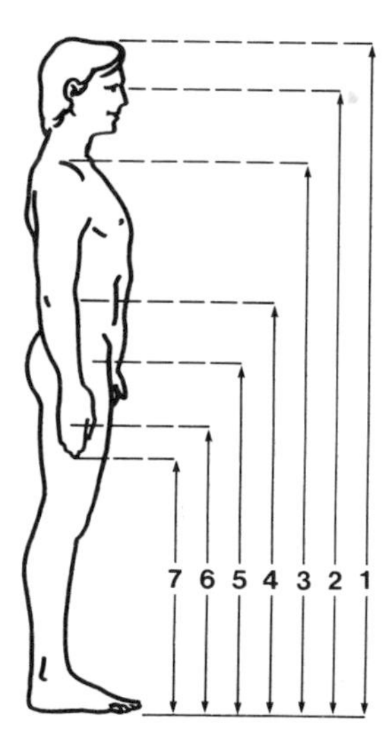

Figure 4.1. Static anthropometric dimensions.

Notes: (*i*) In children under 2 years, who cannot stand unaided, crown–heel length is the nearest equivalent dimension. The child must be stretched out in a supine position and prevented from wriggling.

(*ii*) If you ask an adult person to tell you their height, you must expect them to overestimate by an average of about 25 mm.

2. Eye height

Definition: Vertical distance from the floor to the inner canthus (corner) of the eye.

Applications: Centre of the visual field; reference datum for location of visual displays as in Section 8.4; 'reach' dimension for sight lines, defining maximal acceptable height of visual obstructions; optical sighting devices for prolonged use should be adjustable for the range of users.

Corrections: Shoes as in Section 4.2.

3. Shoulder height

Definition: Vertical distance from the floor to the acromion (i.e., the bony tip of the shoulder).

Applications: The approximate centre of rotation of the upper limb and, hence, of use in determining zones of comfortable reach; reference datum for location of fixtures, fittings, controls, etc. (see Section 7.3).

Correction: Shoes as in Section 4.2.

4. Elbow height

Definition: Vertical distance from the floor to the radiale. (The radiale is the bony landmark formed by the upper end of the radius bone which is palpable on the outer surface of the elbow.)

Applications: An important reference datum for the determination of work-surface heights, etc. (see Sections 8.5 and 15.1).

Correction: Shoes as in Section 4.2.

Note: Some surveys measure to the underside of the elbow when it is flexed to a right angle. This gives a value approximately 15 mm less than the standard measurement.

5. Hip height

Definition: Vertical distance from the floor to the greater trochanter (a bony prominence at the upper end of the thigh bone, palpable on the lateral surface of the hip).

Applications: Centre of rotation of the hip joint, hence the functional length of the lower limb.

Correction: Shoes as in Section 4.2.

6. Knuckle height

Definition: Vertical distance from the floor to metacarpal III (i.e., the knuckle of the middle finger).

Applications: Reference level for handgrips; for support (handrails, etc.) approximately 100 mm above knuckle height is desirable. Handgrips on portable objects should be at less than knuckle height. Optimal height for exertion of lifting force (see Section 14.2).

Correction: Shoes as in Section 4.2.

7. Fingertip height

Definition: Vertical distance from the floor to the dactylion (i.e., the tip of the middle finger).

Application: Lowest acceptable level for finger-operated controls.

Correction: Shoes as in Section 4.2.

8. Sitting height

Definition: Vertical distance from the sitting surface to the vertex (i.e., the crown of the head).

Applications: Clearance required between seat and overhead obstacles.

Corrections: 10 mm for heavy outdoor clothing beneath the buttocks; variable amount for seat compression; 25 mm for a hat; 35 mm for a safety helmet.

9. Sitting eye height

Definition: Vertical distance from the sitting surface to the inner canthus (corner) of the eye.

Applications: See dimension 2.

Corrections: 10 mm for heavy outdoor clothing; up to 40 mm reduction for 'sitting slump'; seat compression.

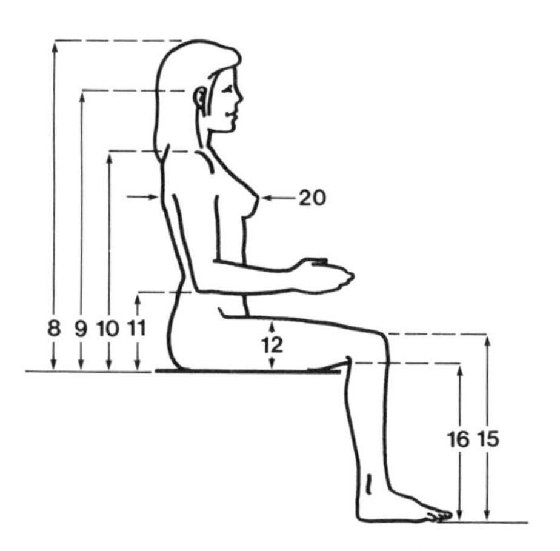

Figure 4.2. Static anthropometric dimensions.

10. Sitting shoulder height

Definition: Vertical distance from the seat surface to the acromion (i.e., the bony point of the shoulder).

Applications: Approximate centre of rotation of the upper limb.

Correction: As for dimension 9.

11. Sitting elbow height (also known as elbow rest height)

Definition: Vertical distance from the seat surface to the underside of the elbow.

Applications: Height of armrests; important reference datum for the heights of desk tops, keyboards, etc., with respect to the seat (see Section 12.1).

12. Thigh thickness (also known as thigh clearance)

Definition: Vertical distance from the seat surface to the top of the uncompressed soft tissue of the thigh at its thickest point, generally where it meets the abdomen.

Applications: Clearance required between the seat and the underside of tables or other obstacles (see Section 12.1).

Correction: 35 mm for heavy outdoor clothing.

13. Buttock–knee length

Definition: Horizontal distance from the back of the uncompressed buttock to the front of the kneecap.

Applications: Clearance between seat back and obstacles in front of the knee (see p. 185).

Correction: 20 mm for heavy outdoor clothing.

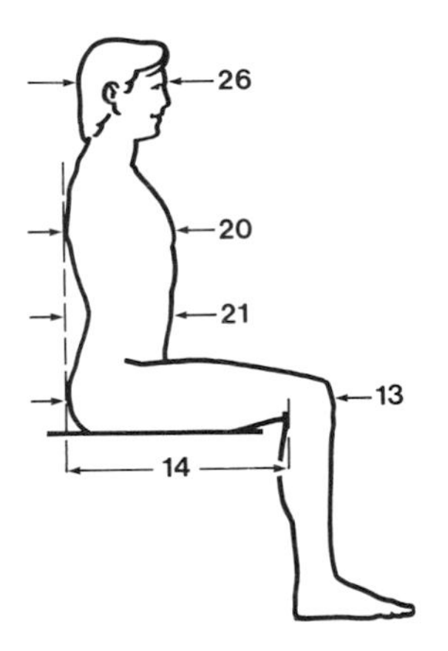

Figure 4.3. Static anthropometric dimensions.

14. Buttock–popliteal length

Definition: Horizontal distance from the back of the uncompressed buttocks to the popliteal angle, at the back of the knee, where the back of the lower legs meet the underside of the thigh.

Applications: Reach dimension, defines maximum acceptable seat depth (see p. 182).

15. Knee height

Definition: Vertical distance from the floor to the upper surface of the knee (usually measured to the quadriceps muscle rather than the kneecap).

Applications: Clearance required beneath the underside of tables, etc. (see p. 184 and Section 12.1).

Corrections: Shoes as in Section 4.2.

16. Popliteal height

Definition: Vertical distance from the floor to the popliteal angle at the underside of the knee where the tendon of the biceps femoris muscle inserts into the lower leg.

Application: Reach dimension defining the maximum acceptable height of a seat (see p. 181 and Section 12.1).

Correction: Shoes as in Section 4.2.

17. Shoulder breadth (bideltoid)

Definition: Maximum horizontal breadth across the shoulders, measured to the protrusions of the deltoid muscles.

Applications: Clearance at shoulder level.

Corrections: 10 mm for indoor clothing; 40 mm for heavy outdoor clothing.

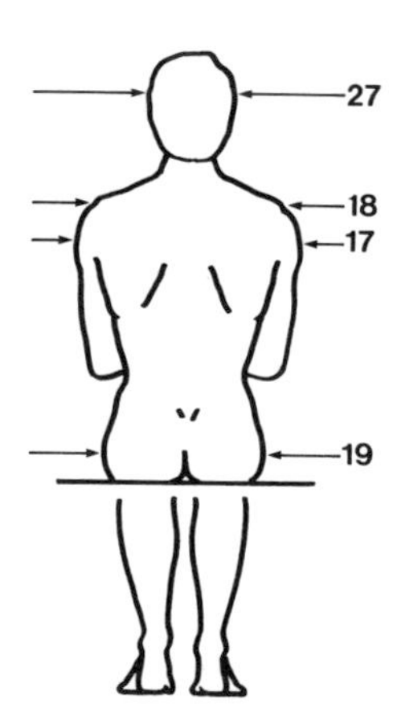

Figure 4.4. Static anthropometric dimensions.

18. Shoulder breadth (biacromial)

Definition: Horizontal distance across the shoulders measured between the acromia (bony points).

Applications: Lateral separation of the centres of rotation of the upper limb.

19. Hip breadth

Definition: Maximum horizontal distance across the hips in the sitting position.

Applications: Clearance at seat level; the width of a seat should be not much less than this (see p. 183).

Corrections: 10 mm for light clothing; 25 mm for medium clothing; 50 mm for heavy outdoor clothing.

Note: This is a dimension with a substantial soft tissue component. In studies of physique, etc., the bony dimension bicristal breadth is generally used (measured between the lateral edges of the crests of the hip bones).

20. Chest (bust) depth

Definition: Maximum horizontal distance from the vertical reference plane to the front of the chest in men or breast in women.

Applications: Clearance between seat backs and obstructions.

Corrections: Up to 40 mm for outdoor clothing.

21. Abdominal depth

Definition: Maximum horizontal distance from the vertical reference plane to the front of the abdomen in the standard sitting position.

Applications: Clearance between seat back and obstructions.

Corrections: Up to 40 mm for outdoor clothing.

22. Shoulder–elbow length

Definition: Distance from the acromion to the underside of the elbow in a standard sitting position.

23. Elbow–fingertip length

Definition: Distance from the back of the elbow to the tip of the middle finger in a standard sitting position.

Applications: Forearm reach; used in defining normal working area (see Section 7.4).

Corrections: For general reach corrections see dimension 34.

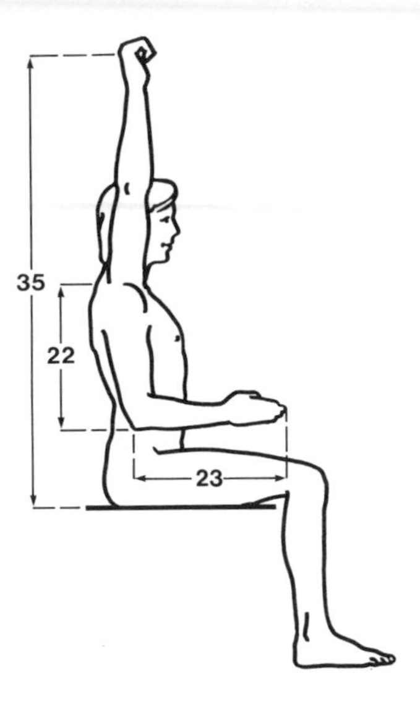

Figure 4.5. Static anthropometric dimensions.

24. Upper limb length

Definition: Distance from the acromion to the fingertip with the elbow and wrist straight (extended).

25. Shoulder–grip length

Definition: Distance from the acromion to the centre of an object gripped in the hand, with the elbow and wrist straight.

Applications: Functional length of upper limb; used in defining zone of convenient reach (see Section 7.3).

Corrections: Reach corrections as in dimension 34.

26. Head length

Definition: Distance between the glabella (the most anterior point on the forehead between the brow ridges) and the occiput (back of the head) in the midline.

Applications: Reference datum for location of eyes, approximately 20 mm behind glabella (see Section 5.1).

27. Head breadth

Definition: Maximum breadth of the head above the level of the ears.

Applications: Clearance.

Corrections: Add 35 mm for clearance across the ears; up to 90 mm for protective helmets.

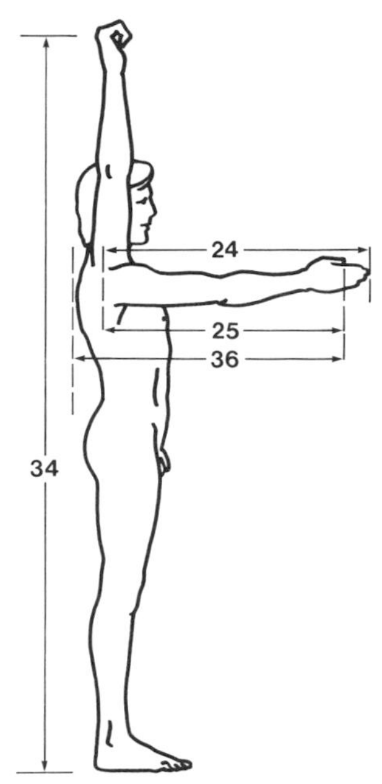

Figure 4.6. Static anthropometric dimensions.

28. Hand length

Definition: Distance from the crease of the wrist to the tip of the middle finger with the hand held straight and stiff.

Applications: See dimension 34 and Section 5.3.

29. Hand breadth

Definition: Maximum breadth across the palm of the hand (at the distal ends of the metacarpal bones).

Applications: Clearance required for hand access, e.g., grips, handles, etc. (see p. 81 and Section 16.1).

Corrections: As much as 25 mm for some protective gloves.

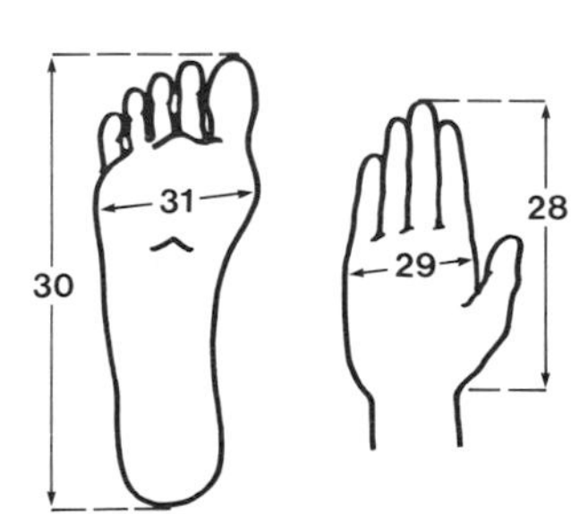

Figure 4:7. Static anthropometric dimensions.

30. Foot length

Definition: Distance, parallel to the long axis of the foot, from the back of the heel to the tip of the longest toe.

Applications: Clearance for foot, design of pedals.

Corrections: In many respects surveys of shoes would be more relevant than surveys of feet since their sizes and shapes are often unrelated. For the purposes of argument we could add 30 mm for men's street shoes and 40 mm for protective boots.

31. Foot breadth

Definition: Maximum horizontal breadth, wherever found, across the foot perpendicular to the long axis.

Applications: Clearance for foot, spacing of pedals, etc.

Corrections: See dimension 30; add 10 mm for men's street shoes; 30 mm for heavy boots.

32. Span

Definition: The maximum horizontal distance between the fingertips when both arms are stretched out sideways.

Application: Lateral reach (see Section 7.1).

Corrections: See dimension 34.

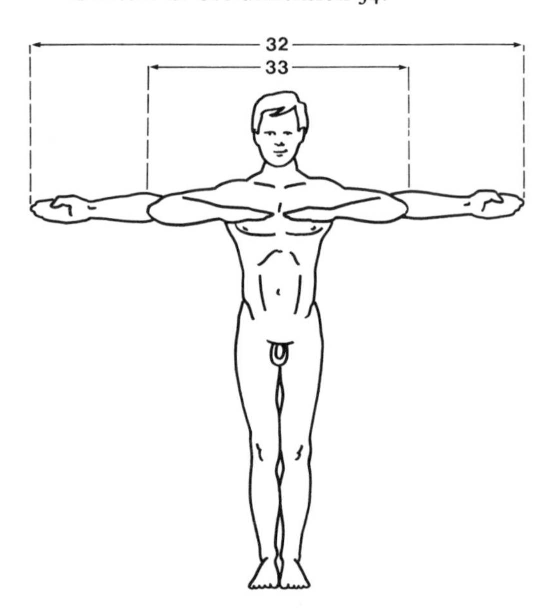

Figure 4.8. Static anthropometric dimensions.

33. Elbow span

Definition: Distance between the tips of the elbows when both upper limbs are stretched out sideways and the elbows are fully flexed so that the fingertips touch the chest.

Applications: A useful guideline when considering 'elbow room' in the workspace.

34–36. Grip reaches

Definitions: In each case the measurement is made to the centre of a cylindrical rod fully grasped in the palm of the hand. In dimensions 34 and 35 the arm is raised vertically above the head and the measurement is made from the floor or seat surface, respectively. In dimension 36 the arm is raised horizontally forward at shoulder level and the measurement is taken from the back of the shoulder blades. In each case these are 'easy' reaches made without excessive stretch.

Corrections: Some surveys measure reach to the tip of the outstretched middle finger or to the tip of the thumb when it forms a 'pinch' with the index finger. Approximately,

$$\text{fingertip reach} = \text{grip reach} + 60\% \text{ hand length}$$

$$\text{thumbtip reach} = \text{grip reach} + 20\% \text{ hand length}$$

See Section 7.2 for a discussion of reach envelopes in general.

4.5. *Notes concerning the data of Tables 4.1–4.37*

Tables 4.1–4.37 have been produced mainly by the method of ratio-scaling described on p. 39 and justified in Section 4.1 and in Pheasant (1982 a). Each table is derived by a combination of two or more sources:

A size source which gives us the mean and standard deviation for stature (or crown–heel length) in the relevant target population.

One or more shape sources from which the coefficients E_1 and E_2 (as defined in Equations 2.45 and 2.46) are calculated. The shape sources must be of the same age and sex as the target population and have a similar ethnic admixture.

Tables 4.2–4.7

Infants

The crown–heel length data were taken from Tanner *et al.* (1966). Male and female data were combined using Equations 2.27 and 2.28 and the 'point in time' values of the original were converted to 'period of time' values using the equation of Healy (1962). Values of E_1 and E_2 were taken from the survey of US infants published by Snyder *et al.* (1977)—linear interpolation was required to adjust the data to appropriate midsample ages.

Tables 4.8–4.23

Children and youths (2–18 years)

In these tables a 5-year-old, for example, is anyone between their fifth and sixth birthdays. The stature data employed were taken from a major survey of British schoolchildren published by the Department of Education and Science (1972). Shape data were from Martin (1960) and Snyder *et al.* (1977). The predictions were edited to ensure steady unidirectional growth in all percentiles and compatibility with young adult data. The data are the same as those of the Department of Education and Science (1985) except that a different selection of anthropometric variables are included—additional details of the editing may be found there. The 2-year-old data presented a particular problem since no suitable sources existed. It was therefore assumed that 2- and 3-year-olds differed by the same amount as 3- and 4-year-olds. No suitable sources exist for the chest or abdominal depths of the under-18-year-olds. E coefficients were calculated from Snyder *et al.* (1977) for equivalent circumferences; they were then scaled down according to young adult depth data. Relative chest depth for girls was assumed to be the same as that for boys until 11 years and then to approach the adult female proportions by steady annual increments.

Tables 4.1 and 4.24–4.28

British adults

Stature data were taken from a survey of a nationwide stratified sample of households conducted in 1981 by the Office of Population Censuses and Surveys (OPCS 1981, Knight 1984). We may have considerable confidence in the validity and reliability of these data. Reference sources for E_1 and E_2 were: US civilians (Stoudt *et al.* 1965, 1970) for dimensions 8, 11, 13, 14, 15, 16, 17, 18 and 19; French drivers (Rebiffe *et al.* 1983) for dimensions 22, 23, 24, 25 and 36; British drivers (Haslegrave 1979) for dimensions 12, 20 and 21. The remaining dimensions were calculated from a variety of US military surveys published in NASA (1978). Separate E coefficients were established for the different age bands.

Assuming the secular trend in young adult stature to have come to a halt by the time the OPCS survey was conducted (see Section 3.4), we would, none the less, anticipate a steady upward drift for some decades to come as its after-effects are felt in the older age groups. Table 4.24 is a fair prediction, in the light of currently available evidence, of the dimensions of British adults around the year AD 2000.

The author has published several previous estimates for British adults. The estimates of Grieve and Pheasant (1982) were based on different stature data, since at the time of writing the OPCS survey had not been published. Those of Pheasant (1982 b, 1984) were based on the same stature data but an earlier set of *E* coefficients which have subsequently been improved. Differences are relatively small but the present tables may be deemed to supersede previous compilations.

The over-65-year-olds presented a problem. The OPCS stature data only extend to 65 years. An alternative source would be the survey by the Institute of Consumer Ergonomics (1983) of the inmates of geriatric institutions. The latter were measured shod but if we subtract a nominal 20 mm for heels we still see that their stature is very much less than that of the 45–65-year-olds in the OPCS sample. In the light of this discrepancy two tables have been prepared. Table 4.27 is based on stature data estimated on the assumption that the decline in stature after 65 years is of a similar magnitude in Great Britain and the USA (as documented by Stoudt *et al.* 1965). Table 4.28 is based on the Institute of Consumer Ergonomics (1983) survey—to which the reader should turn for further information.

Tables 4.29–4.37

Adult populations of other nationalities

Table 4.29 employs stature data from a major survey of US adults conducted in 1971–1974 and deemed to be the most representative and up to date available (Abraham 1979). *E* coefficients were the same as for Table 4.1.

Table 4.30 is based on a survey of French drivers by Rebiffe *et al.* (1983); Table 4.31 on a survey of 25–40-year-old German service men conducted by Jugens *et al.*, cited in NASA (1978), together with data for female office workers from Von Peters (1969); Table 4.32 on Swedish male and female industrial workers from Lewin (1969) and Swedish women from Ingelmark and Lewin (1968); Table 4.33 on Swiss industrial workers from Grandjean (1973); and Table 4.34 on Polish industrial workers from Batogowska and Slowikowski (1974). In each case dimensions marked with an asterisk (*) are quoted directly (when the asterisk is followed by an (M) or a (W), then this applies only to the data for either men or women); the remainder are estimated using the same *E* coefficients a for Table 4.1.

Stature data and *E* coefficients for men used in Table 4.35 are based on sources cited in NASA (1978); *E* coefficients for women based on the assumption that differences in proportion between Japanese men and women are similar to those for European men and women. (In the absence of suitable reference data estimates were made by scaling from the nearest available dimension.)

Table 4.36 is based on data from a hitherto unpublished survey of Chinese industrial workers in Hong Kong, kindly supplied by Bill Evans of the Department of Industrial Engineering at the University of Hong Kong. Dimensions marked with an asterisk (*) are quoted directly; the remainder are estimated as for Table 4.35.

Table 4.37 employs average stature data from Eveleth and Tanner (1976). It assumes that the coefficient of variation for stature is 4% and that body lengths are proportionally similar to Europeans and breadths and depths to Japanese. (These assumptions are highly questionable, but I have been unable to locate any better sources of information.)

Tables 4.38 and 4.39

Body weight data

In Table 4.38 some percentiles are quoted directly from the original surveys and others have been calculated from the parameters of the normal distribution. Since weight is commonly skewed the latter are subject to error (see Section 2.1). The magnitude of these errors is not likely to exceed around 3 kg. The cautious reader could add this figure to both the 5th and 95th %iles. Data are from the sources cited above.

4.6. *The anthropometric tables*

Table 4.1. Anthropometric estimates for British adults aged 19–65 years (all dimensions in millimetres).

Dimension	Men 5th %ile	Men 50th %ile	Men 95th %ile	Men SD	Women 5th %ile	Women 50th %ile	Women 95th %ile	Women SD
1. Stature	1625	1740	1855	70	1505	1610	1710	62
2. Eye height	1515	1630	1745	69	1405	1505	1610	61
3. Shoulder height	1315	1425	1535	66	1215	1310	1405	58
4. Elbow height	1005	1090	1180	52	930	1005	1085	46
5. Hip height	840	920	1000	50	740	810	885	43
6. Knuckle height	690	755	825	41	660	720	780	36
7. Fingertip height	590	655	720	38	560	625	685	38
8. Sitting height	850	910	965	36	795	850	910	35
9. Sitting eye height	735	790	845	35	685	740	795	33
10. Sitting shoulder height	540	595	645	32	505	555	610	31
11. Sitting elbow height	195	245	295	31	185	235	280	29
12. Thigh thickness	135	160	185	15	125	155	180	17
13. Buttock–knee length	540	595	645	31	520	570	620	30
14. Buttock–popliteal length	440	495	550	32	435	480	530	30
15. Knee height	490	545	595	32	455	500	540	27
16. Popliteal height	395	440	490	29	355	400	445	27
17. Shoulder breadth (bideltoid)	420	465	510	28	355	395	435	24
18. Shoulder breadth (biacromial)	365	400	430	20	325	355	385	18
19. Hip breadth	310	360	405	29	310	370	435	38
20. Chest (bust) depth	215	250	285	22	210	250	295	27
21. Abdominal depth	220	270	325	32	205	255	305	30
22. Shoulder–elbow length	330	365	395	20	300	330	360	17
23. Elbow–fingertip length	440	475	510	21	400	430	460	19
24. Upper limb length	720	780	840	36	655	705	760	32
25. Shoulder–grip length	610	665	715	32	555	600	650	29
26. Head length	180	195	205	8	165	180	190	7
27. Head breadth	145	155	165	6	135	145	150	6
28. Hand length	175	190	205	10	160	175	190	9
29. Hand breadth	80	85	95	5	70	75	85	4
30. Foot length	240	265	285	14	215	235	255	12
31. Foot breadth	85	95	110	6	80	90	100	6
32. Span	1655	1790	1925	83	1490	1605	1725	71
33. Elbow span	865	945	1020	47	780	850	920	43
34. Vertical grip reach (standing)	1925	2060	2190	80	1790	1905	2020	71
35. Vertical grip reach (sitting)	1145	1245	1340	60	1060	1150	1235	53
36. Forward grip reach	720	780	835	34	650	705	755	31

See notes on p. 82.

Table 4.2. Anthropometric estimates for newborn infants (all dimensions in millimetres).

Dimension	5th %ile	50th %ile	95th %ile	SD
1. Crown–heel length (1)[a]	465	500	535	20
2. Crown–rump length (8)	330	350	370	11
3. Rump–knee length (13)	105	125	140	10
4. Knee–sole length (15)	120	135	145	7
5. Shoulder breadth (bideltoid) (17)	135	150	160	8
6. Hip breadth (19)	105	120	130	8
7. Shoulder–elbow length (22)	85	95	105	6
8. Elbow–fingertip length (23)	120	135	145	7
9. Head length (26)	115	120	125	4
10. Head breadth (27)	90	95	100	3
11. Hand length (28)	55	60	65	3
12. Hand breadth (29)	30	35	35	2
13. Foot length (30)	65	75	80	4
14. Foot breadth (31)	30	30	35	2

[a]Numbers in parentheses are the equivalent dimensions in adult tables.

See notes on p. 81.

Table 4.3. Anthropometric estimates for infants less than 6 months of age (all dimensions in millimetres).

Dimension	5th %ile	50th %ile	95th %ile	SD
1. Crown–heel length (1)[a]	510	600	690	54
2. Crown–rump length (8)	360	410	460	31
3. Rump–knee length (13)	105	150	195	28
4. Knee–sole length (15)	125	160	190	19
5. Shoulder breadth (bideltoid) (17)	140	180	215	21
6. Hip breadth (19)	100	140	175	22
7. Shoulder–elbow length (22)	90	115	145	16
8. Elbow–fingertip length (23)	125	160	190	19
9. Head length (26)	120	140	160	11
10. Head breadth (27)	100	110	120	7
11. Hand length (28)	55	70	85	9
12. Hand breadth (29)	30	40	45	4
13. Foot length (30)	70	85	105	12
14. Foot breadth (31)	30	40	50	6

[a]Numbers in parentheses are the equivalent dimensions in adult tables.

See notes on p. 81.

Table 4.4. Anthropometric estimates for infants from 6 months to 1 year (all dimensions in millimetres).

Dimension	5th %ile	50th %ile	95th %ile	SD
1. Crown–heel length (1)[a]	655	715	775	37
2. Crown–rump length (8)	435	470	505	21
3. Rump–knee length (13)	155	185	215	19
4. Knee–sole length (15)	170	190	210	13
5. Shoulder breadth (bideltoid) (17)	185	210	230	14
6. Hip breadth (19)	140	165	190	15
7. Shoulder–elbow length (22)	120	140	155	11
8. Elbow–fingertip length (23)	170	190	210	13
9. Head length (26)	145	160	170	8
10. Head breadth (27)	115	120	130	5
11. Hand length (28)	75	85	95	6
12. Hand breadth (29)	40	45	50	3
13. Foot length (30)	90	105	120	8
14. Foot breadth (31)	40	45	50	4

[a]Numbers in parentheses are the equivalent dimensions in adult tables.

See notes on p. 81.

Table 4.5. Anthropometric estimates for infants from 1 year to 18 months (all dimensions in millimetres).

Dimension	5th %ile	50th %ile	95th %ile	SD
1. Crown–heel length (1)[a]	690	745	800	35
2. Crown–rump length (8)	440	475	505	20
3. Rump–knee length (13)	170	195	225	18
4. Knee–sole length (15)	175	195	215	12
5. Shoulder breadth (bideltoid) (17)	185	205	230	14
6. Hip breadth (19)	140	165	185	14
7. Shoulder–elbow length (22)	130	145	160	10
8. Elbow–fingertip length (23)	175	195	215	12
9. Head length (26)	150	160	170	7
10. Head breadth (27)	115	120	130	5
11. Hand length (28)	80	90	100	6
12. Hand breadth (29)	40	45	50	3
13. Foot length (30)	100	115	125	8
14. Foot breadth (31)	40	45	55	4

[a]Numbers in parentheses are the equivalent dimensions in adult tables.

See notes on p. 81.

Table 4.6. Anthropometric estimates for infants from 18 months to 2 years (all dimensions in millimetres).

Dimension	5th %ile	50th %ile	95th %ile	SD
1. Crown–heel length (1)[a]	780	840	900	36
2. Crown–rump length (8)	490	525	555	20
3. Rump–knee length (13)	200	230	260	18
4. Knee–sole length (15)	210	230	255	13
5. Shoulder breadth (bideltoid) (17)	205	230	250	14
6. Hip breadth (19)	150	175	200	15
7. Shoulder–elbow length (22)	145	165	180	11
8. Elbow–fingertip length (23)	200	220	240	12
9. Head length (26)	160	175	185	7
10. Head breadth (27)	125	130	140	5
11. Hand length (28)	85	95	105	6
12. Hand breadth (29)	45	50	55	3
13. Foot length (30)	115	125	140	8
14. Foot breadth (31)	45	55	60	4

[a]Numbers in parentheses are the equivalent dimensions in adult tables.

See notes on p. 81.

Table 4.7. Anthropometric estimates for British 2-year-olds (all dimensions in millimetres).

Dimension	Boys				Girls			
	5th %ile	50th %ile	95th %ile	SD	5th %ile	50th %ile	95th %ile	SD
1. Stature	850	930	1010	49	825	890	955	40
2. Eye height	760	840	920	49	725	805	885	48
3. Shoulder height	675	735	795	37	630	695	760	38
4. Elbow height	495	555	615	36	480	530	580	30
5. Hip height	360	420	480	37	365	415	465	30
6. Knuckle height	340	385	430	26	335	375	415	25
7. Fingertip height	275	315	360	26	270	310	350	25
8. Sitting height	505	545	585	24	485	520	555	21
9. Sitting eye height	410	445	480	20	370	410	450	24
10. Sitting shoulder height	305	340	375	22	275	310	345	20
11. Sitting elbow height	105	140	175	20	105	130	155	15
12. Thigh thickness	65	80	95	10	60	75	90	10
13. Buttock–knee length	245	275	305	19	250	280	310	17
14. Buttock–popliteal length	210	235	260	16	185	245	305	36
15. Knee height	235	270	305	20	230	260	290	17
16. Popliteal height	155	205	255	29	170	205	240	20
17. Shoulder breadth (bideltoid)	215	245	275	17	210	235	260	14
18. Shoulder breadth (biacromial)	190	215	240	15	190	210	230	12
19. Hip breadth	170	190	210	13	165	185	205	11
20. Chest (bust) depth	100	120	140	12	100	115	130	10
21. Abdominal depth	130	145	160	10	135	145	155	7
22. Shoulder–elbow length	160	185	205	13	160	175	190	10
23. Elbow–fingertip length	215	245	275	17	210	235	260	14
24. Upper limb length	365	410	455	28	335	380	425	27
25. Shoulder–grip length	295	340	390	28	270	315	360	27
26. Head length	170	180	190	7	160	165	170	4
27. Head breadth	130	140	150	6	125	130	135	4
28. Hand length	90	105	120	8	90	100	110	6
29. Hand breadth	50	55	60	4	40	45	50	4
30. Foot length	130	145	160	10	130	145	160	9
31. Foot breadth	60	65	70	4	50	55	60	4
32. Span	835	925	1015	54	785	865	945	49
33. Elbow span	435	490	540	31	410	455	505	30
34. Vertical grip reach (standing)	920	1045	1170	77	965	1045	1125	49
35. Vertical grip reach (sitting)	605	675	745	42	550	620	690	42
36. Forward grip reach	340	400	460	35	345	385	425	25

See notes on p. 81.

Table 4.8. Anthropometric estimates for British 3-year-olds (all dimensions in millimetres).

Dimension	Boys				Girls			
	5th %ile	50th %ile	95th %ile	SD	5th %ile	50th %ile	95th %ile	SD
1. Stature	910	990	1070	48	895	970	1045	46
2. Eye height	810	890	970	48	785	875	965	55
3. Shoulder height	720	780	840	36	690	760	830	43
4. Elbow height	535	595	655	35	520	580	640	35
5. Hip height	400	460	520	35	405	460	515	33
6. Knuckle height	365	410	455	26	360	410	460	29
7. Fingertip height	295	340	380	26	290	340	385	29
8. Sitting height	530	570	610	25	515	555	595	25
9. Sitting eye height	425	465	505	24	400	445	490	28
10. Sitting shoulder height	310	350	390	23	295	335	375	23
11. Sitting elbow height	115	150	185	20	110	140	170	17
12. Thigh thickness	65	85	105	11	60	80	100	12
13. Buttock–knee length	270	300	330	19	270	305	340	20
14. Buttock–popliteal length	225	250	275	16	215	260	305	26
15. Knee height	255	290	325	20	250	285	320	20
16. Popliteal height	195	230	265	21	200	230	260	17
17. Shoulder breadth (bideltoid)	230	255	280	16	225	250	275	15
18. Shoulder breadth (biacromial)	200	225	250	14	205	225	245	13
19. Hip breadth	175	195	215	13	175	195	215	13
20. Chest (bust) depth	105	125	145	12	105	120	140	12
21. Abdominal depth	135	150	165	10	135	150	165	10
22. Shoulder–elbow length	175	195	220	13	175	195	215	12
23. Elbow–fingertip length	235	260	285	16	230	255	280	16
24. Upper limb length	390	435	480	27	365	415	465	31
25. Shoulder–grip length	320	365	410	27	295	345	395	31
26. Head length	170	180	190	7	155	165	175	6
27. Head breadth	130	140	150	6	120	130	140	5
28. Hand length	95	110	125	8	100	110	120	7
29. Hand breadth	50	55	60	4	45	50	55	4
30. Foot length	140	155	170	10	140	155	170	10
31. Foot breadth	60	65	70	4	55	60	65	4
32. Span	890	980	1070	56	850	940	1030	56
33. Elbow span	465	515	570	32	440	495	555	34
34. Vertical grip reach (standing)	1005	1130	1255	75	1025	1125	1225	61
35. Vertical grip reach (sitting)	640	705	775	41	605	675	740	41
36. Forward grip reach	360	420	480	35	360	415	470	33

See notes on p. 82.

Table 4.9. Anthropometric estimates for British 4-year-olds (all dimensions in millimetres).

Dimension	Boys				Girls			
	5th %ile	50th %ile	95th %ile	SD	5th %ile	50th %ile	95th %ile	SD
1. Stature	975	1050	1125	47	965	1050	1135	52
2. Eye height	865	940	1015	47	845	945	1045	62
3. Shoulder height	765	825	885	35	745	825	905	48
4. Elbow height	580	635	690	34	565	630	695	40
5. Hip height	445	500	555	33	445	505	565	36
6. Knuckle height	390	435	480	26	390	445	500	33
7. Fingertip height	315	360	405	26	315	365	420	33
8. Sitting height	550	595	640	26	540	590	640	29
9. Sitting eye height	440	485	530	28	425	480	535	32
10. Sitting shoulder height	320	360	400	24	315	360	405	26
11. Sitting elbow height	125	160	195	20	120	150	180	19
12. Thigh thickness	70	90	110	12	60	85	110	14
13. Buttock–knee length	295	325	355	19	290	330	370	23
14. Buttock–popliteal length	240	265	290	16	250	275	300	16
15. Knee height	275	310	345	20	270	310	350	23
16. Popliteal height	235	255	275	13	230	255	280	14
17. Shoulder breadth (bideltoid)	240	265	290	15	240	265	290	16
18. Shoulder breadth (biacromial)	215	235	255	13	215	240	265	14
19. Hip breadth	180	200	220	13	180	205	230	15
20. Chest (bust) depth	110	130	150	12	110	130	150	13
21. Abdominal depth	140	155	170	10	135	155	175	13
22. Shoulder–elbow length	190	210	230	13	185	210	235	14
23. Elbow–fingertip length	250	275	300	15	245	275	305	18
24. Upper limb length	415	460	505	26	390	450	510	35
25. Shoulder–grip length	340	385	430	26	315	370	430	35
26. Head length	170	180	190	7	150	165	180	8
27. Head breadth	130	140	150	6	125	135	145	6
28. Hand length	100	115	130	8	105	120	135	8
29. Hand breadth	50	55	60	4	50	55	60	4
30. Foot length	150	165	180	10	145	165	185	11
31. Foot breadth	60	65	70	4	60	65	70	4
32. Span	940	1035	1130	58	910	1015	1120	63
33. Elbow span	490	545	600	33	475	535	600	38
34. Vertical grip reach (standing)	1095	1215	1335	73	1085	1205	1325	73
35. Vertical grip reach (sitting)	670	735	805	40	660	725	795	40
36. Forward grip reach	380	440	500	35	380	445	510	41

See notes on p. 82.

Table 4.10. Anthropometric estimates for British 5-year-olds (all dimensions in millimetres).

Dimension	Boys				Girls			
	5th %ile	50th %ile	95th %ile	SD	5th %ile	50th %ile	95th %ile	SD
1. Stature	1025	1110	1195	52	1015	1100	1185	53
2. Eye height	910	995	1080	53	885	990	1095	64
3. Shoulder height	810	875	940	39	785	865	945	48
4. Elbow height	605	670	735	39	595	660	725	41
5. Hip height	490	550	610	36	490	540	590	31
6. Knuckle height	405	455	505	29	410	465	520	34
7. Fingertip height	325	375	420	29	330	385	445	34
8. Sitting height	575	620	665	28	560	610	660	29
9. Sitting eye height	455	505	555	29	450	500	550	31
10. Sitting shoulder height	340	380	420	25	325	370	415	26
11. Sitting elbow height	130	165	200	22	125	155	185	19
12. Thigh thickness	75	90	105	10	70	90	110	12
13. Buttock–knee length	310	345	380	21	310	350	390	23
14. Buttock–popliteal length	250	280	310	17	265	295	325	19
15. Knee height	300	335	370	22	295	330	365	21
16. Popliteal height	240	270	300	18	245	270	295	16
17. Shoulder breadth (bideltoid)	245	275	305	17	245	270	295	16
18. Shoulder breadth (biacromial)	225	250	275	15	230	250	270	12
19. Hip breadth	185	210	235	15	185	210	235	16
20. Chest (bust) depth	110	135	160	14	110	135	155	14
21. Abdominal depth	135	155	175	12	135	160	185	14
22. Shoulder–elbow length	205	225	250	14	200	220	245	13
23. Elbow–fingertip length	265	295	325	17	260	290	320	17
24. Upper limb length	435	485	535	29	410	470	530	35
25. Shoulder–grip length	355	405	450	29	335	390	450	35
26. Head length	165	180	195	8	150	165	180	8
27. Head breadth	130	140	150	5	120	130	140	5
28. Hand length	110	125	140	9	105	120	135	8
29. Hand breadth	55	60	65	4	50	55	60	4
30. Foot length	155	175	195	11	155	170	185	10
31. Foot breadth	60	70	80	5	60	65	70	4
32. Span	995	1095	1195	60	955	1060	1165	64
33. Elbow span	520	575	635	34	495	560	625	39
34. Vertical grip reach (standing)	1180	1305	1430	77	1170	1290	1410	72
35. Vertical grip reach (sitting)	700	775	850	45	685	755	830	45
36. Forward grip reach	420	470	520	30	400	460	520	36

See notes on p. 82.

Table 4.11. Anthropometric estimates for British 6-year-olds (all dimensions in millimetres).

	Boys				Girls			
Dimension	5th %ile	50th %ile	95th %ile	SD	5th %ile	50th %ile	95th %ile	SD
1. Stature	1070	1170	1270	60	1070	1160	1250	56
2. Eye height	950	1050	1150	60	935	1045	1155	67
3. Shoulder height	845	920	995	45	825	910	995	52
4. Elbow height	635	705	775	44	625	695	765	43
5. Hip height	520	595	670	45	420	475	530	32
6. Knuckle height	425	480	535	33	430	490	550	36
7. Fingertip height	340	395	450	33	350	410	470	36
8. Sitting height	585	640	695	32	585	635	685	31
9. Sitting eye height	475	525	575	31	470	525	580	32
10. Sitting shoulder height	340	390	440	29	335	380	425	28
11. Sitting elbow height	130	170	210	25	125	160	195	21
12. Thigh thickness	75	95	115	13	75	95	115	11
13. Buttock–knee length	330	370	410	25	330	370	410	25
14. Buttock–popliteal length	270	305	340	21	275	310	345	20
15. Knee height	320	360	400	25	320	355	390	21
16. Popliteal height	260	295	330	22	265	290	315	16
17. Shoulder breadth (bideltoid)	245	285	325	23	250	285	320	20
18. Shoulder breadth (biacromial)	235	265	295	18	240	260	280	13
19. Hip breadth	180	215	250	21	190	220	250	19
20. Chest (bust) depth	110	140	170	19	110	140	170	18
21. Abdominal depth	135	160	185	16	135	165	195	18
22. Shoulder–elbow length	215	240	265	16	215	235	255	13
23. Elbow–fingertip length	275	310	345	21	275	305	335	18
24. Upper limb length	455	510	565	34	430	495	560	38
25. Shoulder–grip length	370	425	480	34	350	415	475	38
26. Head length	165	180	195	9	160	170	180	7
27. Head breadth	130	140	150	6	125	135	145	6
28. Hand length	115	130	145	10	110	125	140	8
29. Hand breadth	50	60	70	5	55	60	65	4
30. Foot length	165	185	205	13	160	180	200	11
31. Foot breadth	65	75	85	6	60	70	80	5
32. Span	1045	1160	1275	70	1010	1120	1230	68
33. Elbow span	545	610	675	40	525	590	660	41
34. Vertical grip reach (standing)	1235	1390	1545	93	1255	1380	1505	76
35. Vertical grip reach (sitting)	720	805	890	52	705	790	875	52
36. Forward grip reach	435	495	555	35	435	485	535	31

See notes on p. 82.

Table 4.12. Anthropometric estimates for British 7-year-olds (all dimensions in millimetres).

Dimension	Boys				Girls			
	5th %ile	50th %ile	95th %ile	SD	5th %ile	50th %ile	95th %ile	SD
1. Stature	1140	1230	1320	56	1125	1220	1315	59
2. Eye height	1020	1115	1210	57	995	1105	1215	66
3. Shoulder height	885	975	1065	54	870	960	1050	54
4. Elbow height	680	745	810	40	665	735	805	42
5. Hip height	570	635	700	39	555	615	675	35
6. Knuckle height	460	510	560	31	465	525	585	37
7. Fingertip height	370	420	475	31	375	435	500	37
8. Sitting height	615	665	715	30	610	660	710	31
9. Sitting eye height	505	550	595	28	500	555	610	32
10. Sitting shoulder height	360	405	450	27	350	395	440	26
11. Sitting elbow height	140	175	210	20	140	170	200	19
12. Thigh thickness	85	105	125	13	85	105	125	13
13. Buttock–knee length	355	395	435	24	355	400	445	26
14. Buttock–popliteal length	280	325	370	27	290	335	380	27
15. Knee height	340	380	420	25	335	375	415	23
16. Popliteal height	285	315	345	19	275	310	345	21
17. Shoulder breadth (bideltoid)	265	300	335	22	255	295	335	24
18. Shoulder breadth (biacromial)	250	275	300	15	245	270	295	15
19. Hip breadth	190	225	260	21	195	235	275	23
20. Chest (bust) depth	110	145	180	20	110	145	180	21
21. Abdominal depth	135	165	195	19	130	170	210	23
22. Shoulder–elbow length	230	255	280	15	225	250	275	15
23. Elbow–fingertip length	295	325	355	19	290	320	350	18
24. Upper limb length	485	540	595	32	470	525	580	34
25. Shoulder–grip length	400	450	505	32	380	435	495	34
26. Head length	170	185	200	8	160	170	180	6
27. Head breadth	130	140	150	5	125	135	145	5
28. Hand length	120	135	150	9	120	135	150	8
29. Hand breadth	60	65	70	4	55	60	65	4
30. Foot length	175	195	215	11	170	190	210	12
31. Foot breadth	65	75	85	5	65	75	85	5
32. Span	1125	1230	1335	64	1095	1195	1295	62
33. Elbow span	590	650	710	36	570	630	695	38
34. Vertical grip reach (standing)	1350	1475	1600	76	1325	1455	1585	79
35. Vertical grip reach (sitting)	770	850	925	48	745	825	905	48
36. Forward grip reach	470	520	570	31	455	505	555	29

See notes on p. 82.

Table 4.13. Anthropometric estimates for British 8-year-olds (all dimensions in millimetres).

Dimension	Boys				Girls			
	5th %ile	50th %ile	95th %ile	SD	5th %ile	50th %ile	95th %ile	SD
1. Stature	1180	1280	1380	60	1185	1280	1375	59
2. Eye height	1070	1165	1260	59	1070	1165	1260	58
3. Shoulder height	930	1020	1110	54	930	1015	1100	53
4. Elbow height	705	780	855	45	705	775	845	42
5. Hip height	605	665	725	35	585	650	715	38
6. Knuckle height	480	535	590	32	495	555	615	37
7. Fingertip height	390	445	495	32	405	465	525	37
8. Sitting height	630	680	730	31	640	685	730	28
9. Sitting eye height	520	570	620	31	525	580	635	32
10. Sitting shoulder height	380	425	470	27	370	410	450	25
11. Sitting elbow height	145	180	215	21	145	175	205	19
12. Thigh thickness	85	110	135	14	90	110	130	13
13. Buttock–knee length	375	415	455	25	375	420	465	26
14. Buttock–popliteal length	305	340	375	22	310	355	400	27
15. Knee height	360	400	440	25	355	395	435	24
16. Popliteal height	295	325	355	18	295	330	365	20
17. Shoulder breadth (bideltoid)	275	310	345	21	270	310	350	24
18. Shoulder breadth (biacromial)	265	285	305	13	255	280	305	16
19. Hip breadth	200	235	270	20	205	245	285	23
20. Chest (bust) depth	115	150	185	20	120	150	180	20
21. Abdominal depth	135	170	205	20	140	180	220	24
22. Shoulder–elbow length	240	265	290	15	240	260	285	14
23. Elbow–fingertip length	310	340	370	19	305	335	365	19
24. Upper limb length	515	565	615	30	495	555	615	35
25. Shoulder–grip length	425	475	525	30	405	465	520	35
26. Head length	170	185	200	8	165	175	185	5
27. Head breadth	130	140	150	5	125	135	145	5
28. Hand length	125	140	155	9	125	140	155	8
29. Hand breadth	60	65	70	4	60	65	70	4
30. Foot length	180	200	220	12	180	200	220	12
31. Foot breadth	70	80	90	5	65	75	85	5
32. Span	1165	1280	1395	69	1150	1250	1350	60
33. Elbow span	610	675	740	39	600	660	720	36
34. Vertical grip reach (standing)	1425	1550	1675	75	1405	1535	1665	78
35. Vertical grip reach (sitting)	805	890	975	52	785	870	955	52
36. Forward grip reach	475	535	595	35	475	530	585	34

See notes on p. 82.

Table 4.14. Anthropometric estimates for British 9-year-olds (all dimensions in millimetres).

Dimension	Boys				Girls			
	5th %ile	50th %ile	95th %ile	SD	5th %ile	50th %ile	95th %ile	SD
1. Stature	1225	1330	1435	63	1220	1330	1440	68
2. Eye height	1005	1110	1215	64	1105	1215	1325	67
3. Shoulder height	965	1065	1165	60	955	1060	1165	63
4. Elbow height	740	820	900	50	720	815	910	57
5. Hip height	635	700	765	40	610	690	770	48
6. Knuckle height	505	565	625	36	530	590	650	37
7. Fingertip height	410	470	530	36	435	495	555	37
8. Sitting height	650	700	750	31	645	700	755	33
9. Sitting eye height	530	585	640	33	540	595	650	33
10. Sitting shoulder height	390	440	490	29	385	430	475	28
11. Sitting elbow height	150	190	230	24	140	180	220	25
12. Thigh thickness	90	115	140	15	90	115	140	15
13. Buttock–knee length	395	440	485	26	395	445	495	30
14. Buttock–popliteal length	325	365	405	25	330	380	430	31
15. Knee height	375	420	465	27	375	420	465	27
16. Popliteal height	300	340	380	23	300	340	380	24
17. Shoulder breadth (bideltoid)	280	320	360	23	285	320	355	20
18. Shoulder breadth (biacromial)	270	295	320	15	265	295	325	19
19. Hip breadth	205	245	285	24	210	255	300	27
20. Chest (bust) depth	120	155	190	22	115	155	195	24
21. Abdominal depth	145	180	215	21	140	185	230	26
22. Shoulder–elbow length	250	275	305	16	245	275	300	17
23. Elbow–fingertip length	320	355	390	21	310	350	390	23
24. Upper limb length	530	585	640	33	500	575	650	45
25. Shoulder–grip length	435	490	545	33	405	480	555	45
26. Head length	170	185	200	8	165	175	185	7
27. Head breadth	135	145	155	5	125	135	145	6
28. Hand length	130	145	160	9	130	145	160	10
29. Hand breadth	60	65	70	4	60	65	70	4
30. Foot length	185	210	235	14	185	210	235	14
31. Foot breadth	70	80	90	5	70	80	90	6
32. Span	1200	1330	1460	78	1180	1300	1420	74
33. Elbow span	630	700	775	44	615	685	760	45
34. Vertical grip reach (standing)	1475	1610	1745	83	1460	1615	1770	94
35. Vertical grip reach (sitting)	830	920	1010	54	815	905	995	54
36. Forward grip reach	495	555	615	36	485	555	625	42

See notes on p. 82.

Table 4.15. Anthropometric estimates for British 10-year-olds (all dimensions in millimetres).

Dimension	Boys				Girls			
	5th %ile	50th %ile	95th %ile	SD	5th %ile	50th %ile	95th %ile	SD
1. Stature	1290	1390	1490	61	1270	1390	1510	72
2. Eye height	1180	1275	1370	58	1155	1275	1395	72
3. Shoulder height	1025	1120	1215	57	1015	1120	1225	65
4. Elbow height	770	860	950	55	765	860	955	57
5. Hip height	660	735	810	46	650	730	810	50
6. Knuckle height	540	595	650	33	555	615	675	36
7. Fingertip height	445	500	550	33	460	520	575	36
8. Sitting height	670	725	780	32	665	725	785	36
9. Sitting eye height	550	600	650	29	555	615	675	35
10. Sitting shoulder height	410	455	500	28	400	450	500	30
11. Sitting elbow height	160	195	230	21	150	190	230	25
12. Thigh thickness	100	120	140	13	95	120	145	16
13. Buttock–knee length	415	460	505	27	415	470	525	32
14. Buttock–popliteal length	340	380	420	25	350	400	450	29
15. Knee height	395	440	485	26	395	440	485	28
16. Popliteal height	330	360	390	19	325	365	405	25
17. Shoulder breadth (bideltoid)	290	335	380	27	280	330	380	31
18. Shoulder breadth (biacromial)	275	305	335	18	275	305	335	19
19. Hip breadth	215	260	305	28	215	265	315	30
20. Chest (bust) depth	120	165	210	26	115	165	215	31
21. Abdominal depth	145	185	225	25	145	190	235	27
22. Shoulder–elbow length	265	290	315	16	260	290	320	18
23. Elbow–fingertip length	335	370	405	22	330	370	410	25
24. Upper limb length	540	610	680	42	520	590	660	44
25. Shoulder–grip length	445	515	580	42	420	495	565	44
26. Head length	170	185	200	8	160	170	180	7
27. Head breadth	135	145	155	5	125	135	145	5
28. Hand length	135	150	165	9	135	150	165	10
29. Hand breadth	65	70	75	4	60	70	80	5
30. Foot length	195	220	245	14	190	215	240	14
31. Foot breadth	70	85	95	5	70	80	90	7
32. Span	1275	1395	1515	73	1240	1365	1490	77
33. Elbow span	665	735	805	41	645	720	800	47
34. Vertical grip reach (standing)	1540	1680	1820	86	1540	1705	1870	101
35. Vertical grip reach (sitting)	870	955	1045	52	850	935	1020	52
36. Forward grip reach	525	580	635	33	520	585	650	40

See notes on p. 82.

Table 4.16. Anthropometric estimates for British 11-year-olds (all dimensions in millimetres).

Dimension	Boys				Girls			
	5th %ile	50th %ile	95th %ile	SD	5th %ile	50th %ile	95th %ile	SD
1. Stature	1325	1430	1535	65	1310	1440	1570	79
2. Eye height	1215	1315	1415	62	1195	1325	1455	78
3. Shoulder height	1060	1160	1260	60	1050	1165	1280	69
4. Elbow height	795	890	985	57	800	890	980	56
5. Hip height	685	765	845	50	670	750	830	48
6. Knuckle height	560	620	680	35	575	645	715	42
7. Fingertip height	460	520	575	35	475	545	615	42
8. Sitting height	685	740	795	34	680	745	810	41
9. Sitting eye height	575	620	665	28	570	635	700	39
10. Sitting shoulder height	425	470	515	26	415	470	525	33
11. Sitting elbow height	160	200	240	24	155	200	245	26
12. Thigh thickness	100	120	140	11	100	125	150	16
13. Buttock–knee length	435	480	525	28	430	490	550	37
14. Buttock–popliteal length	345	395	445	30	365	410	455	26
15. Knee height	420	460	500	25	405	455	505	30
16. Popliteal height	330	375	420	26	335	375	415	24
17. Shoulder breadth (bideltoid)	300	345	390	26	285	340	395	34
18. Shoulder breadth (biacromial)	280	315	350	21	280	315	350	21
19. Hip breadth	220	265	310	27	225	280	335	34
20. Chest (bust) depth	130	170	210	24	115	175	240	38
21. Abdominal depth	150	190	230	23	145	195	245	29
22. Shoulder–elbow length	270	300	325	16	265	300	330	20
23. Elbow–fingertip length	350	385	420	22	340	385	430	28
24. Upper limb length	560	630	700	43	555	630	705	46
25. Shoulder–grip length	460	530	600	43	455	530	605	46
26. Head length	170	185	200	8	155	170	185	8
27. Head breadth	135	145	155	5	125	135	145	5
28. Hand length	140	155	170	10	135	155	175	11
29. Hand breadth	60	70	80	5	60	70	80	5
30. Foot length	205	225	245	13	195	220	245	14
31. Foot breadth	75	85	95	7	75	85	95	7
32. Span	1310	1440	1570	78	1270	1415	1560	87
33. Elbow span	685	760	830	44	660	750	835	53
34. Vertical grip reach (standing)	1575	1740	1905	100	1575	1760	1945	111
35. Vertical grip reach (sitting)	895	990	1080	56	900	990	1085	56
36. Forward grip reach	535	595	655	37	530	600	670	42

See notes on p. 82.

Table 4.17. Anthropometric estimates for British 12-year-olds (all dimensions in millimetres).

	Boys				Girls			
Dimension	5th %ile	50th %ile	95th %ile	SD	5th %ile	50th %ile	95th %ile	SD
1. Stature	1360	1490	1620	78	1370	1500	1630	79
2. Eye height	1245	1375	1505	78	1255	1385	1515	80
3. Shoulder height	1095	1215	1335	72	1100	1215	1330	69
4. Elbow height	840	930	1020	55	840	940	1040	60
5. Hip height	720	805	890	53	705	780	855	47
6. Knuckle height	580	645	710	40	590	665	740	46
7. Fingertip height	470	540	605	40	480	560	635	46
8. Sitting height	700	765	830	39	700	775	850	45
9. Sitting eye height	590	650	710	37	600	665	730	40
10. Sitting shoulder height	440	490	540	30	435	490	545	32
11. Sitting elbow height	160	205	250	27	155	205	255	31
12. Thigh thickness	105	125	145	13	100	130	160	17
13. Buttock–knee length	445	500	555	32	450	510	570	36
14. Buttock–popliteal length	375	415	455	23	380	435	490	33
15. Knee height	430	480	530	30	420	470	520	29
16. Popliteal height	350	390	430	23	345	385	425	24
17. Shoulder breadth (bideltoid)	315	355	395	25	305	355	405	29
18. Shoulder breadth (biacromial)	290	325	360	21	290	325	360	21
19. Hip breadth	230	275	320	26	235	295	355	35
20. Chest (bust) depth	135	175	215	24	135	190	240	33
21. Abdominal depth	165	200	235	22	155	200	245	27
22. Shoulder–elbow length	280	310	340	18	280	315	345	20
23. Elbow–fingertip length	360	400	440	25	355	400	445	27
24. Upper limb length	600	665	730	41	575	660	745	52
25. Shoulder–grip length	490	560	625	41	465	555	640	52
26. Head length	170	185	200	8	165	175	185	7
27. Head breadth	135	145	155	5	130	140	150	6
28. Hand length	150	165	180	10	145	165	185	11
29. Hand breadth	65	75	85	5	60	70	80	5
30. Foot length	215	235	255	13	205	230	255	14
31. Foot breadth	80	90	100	7	75	85	95	7
32. Span	1355	1510	1665	93	1320	1480	1640	96
33. Elbow span	710	795	885	53	685	780	880	58
34. Vertical grip reach (standing)	1655	1835	2015	110	1650	1835	2020	112
35. Vertical grip reach (sitting)	925	1035	1145	67	925	1035	1145	67
36. Forward grip reach	550	620	690	42	550	625	700	45

See notes on p. 82.

Table 4.18. Anthropometric estimates for British 13-year-olds (all dimensions in millimetres).

	Boys				Girls			
Dimension	5th %ile	50th %ile	95th %ile	SD	5th %ile	50th %ile	95th %ile	SD
1. Stature	1400	1550	1700	91	1430	1550	1670	73
2. Eye height	1285	1435	1585	90	1315	1435	1555	74
3. Shoulder height	1130	1265	1400	81	1145	1255	1365	68
4. Elbow height	870	970	1070	61	875	970	1065	57
5. Hip height	740	835	930	57	725	805	885	50
6. Knuckle height	600	670	740	43	605	675	745	43
7. Fingertip height	490	560	630	43	495	565	635	43
8. Sitting height	710	790	870	49	740	805	870	41
9. Sitting eye height	605	680	755	47	630	695	760	39
10. Sitting shoulder height	450	510	570	37	455	510	565	34
11. Sitting elbow height	165	210	255	28	155	210	265	34
12. Thigh thickness	105	130	155	15	110	135	160	15
13. Buttock–knee length	465	525	585	35	480	530	580	31
14. Buttock–popliteal length	375	435	495	35	400	445	490	27
15. Knee height	440	500	560	35	440	485	530	27
16. Popliteal height	355	405	455	30	350	390	430	25
17. Shoulder breadth (bideltoid)	325	375	425	29	325	370	415	26
18. Shoulder breadth (biacromial)	295	335	375	24	300	335	370	21
19. Hip breadth	245	290	335	28	265	315	365	30
20. Chest (bust) depth	135	185	235	29	150	200	245	29
21. Abdominal depth	165	205	245	24	170	210	250	24
22. Shoulder–elbow length	290	325	360	22	295	325	355	18
23. Elbow–fingertip length	370	420	470	29	375	410	445	22
24. Upper limb length	620	695	770	47	605	680	755	47
25. Shoulder–grip length	505	585	660	47	490	570	645	47
26. Head length	175	190	205	8	165	175	185	6
27. Head breadth	140	150	160	5	130	140	150	5
28. Hand length	150	170	190	12	155	170	185	10
29. Hand breadth	70	80	90	6	70	75	80	4
30. Foot length	220	245	270	16	210	230	250	13
31. Foot breadth	80	90	100	7	80	90	100	6
32. Span	1400	1580	1760	110	1385	1540	1695	93
33. Elbow span	730	835	935	62	720	815	905	56
34. Vertical grip reach (standing)	1720	1905	2090	112	1700	1890	2080	114
35. Vertical grip reach (sitting)	955	1080	1210	78	945	1070	1200	78
36. Forward grip reach	575	655	735	48	575	640	705	41

See notes on p. 82.

Table 4.19. Anthropometric estimates for British 14-year-olds (all dimensions in millimetres).

Dimension	Boys				Girls			
	5th %ile	50th %ile	95th %ile	SD	5th %ile	50th %ile	95th %ile	SD
1. Stature	1480	1630	1780	90	1480	1590	1700	66
2. Eye height	1360	1510	1660	91	1365	1475	1585	66
3. Shoulder height	1205	1335	1465	80	1190	1295	1400	64
4. Elbow height	915	1015	1115	60	900	985	1070	53
5. Hip height	795	870	945	46	735	810	885	45
6. Knuckle height	630	700	770	44	640	705	770	40
7. Fingertip height	510	585	655	44	530	595	660	40
8. Sitting height	750	835	920	52	770	830	890	36
9. Sitting eye height	640	720	800	48	660	720	780	37
10. Sitting shoulder height	470	535	600	39	470	525	580	32
11. Sitting elbow height	165	215	265	30	165	220	275	33
12. Thigh thickness	115	140	165	16	115	140	165	14
13. Buttock–knee length	495	550	605	34	495	545	595	29
14. Buttock–popliteal length	405	460	515	33	415	455	495	25
15. Knee height	465	520	575	33	450	495	540	27
16. Popliteal height	380	425	470	28	355	395	435	25
17. Shoulder breadth (bideltoid)	345	395	445	29	345	385	425	25
18. Shoulder breadth (biacromial)	320	355	390	22	315	345	375	19
19. Hip breadth	260	305	350	28	285	330	375	26
20. Chest (bust) depth	145	195	245	29	165	210	255	27
21. Abdominal depth	175	215	255	23	175	215	255	23
22. Shoulder–elbow length	310	345	380	21	305	335	360	17
23. Elbow–fingertip length	400	445	490	28	385	420	455	21
24. Upper limb length	660	735	810	47	640	700	760	36
25. Shoulder–grip length	540	620	695	47	530	590	650	36
26. Head length	180	190	200	7	165	175	185	6
27. Head breadth	140	150	160	6	130	140	150	5
28. Hand length	160	180	200	11	155	170	185	9
29. Hand breadth	75	85	95	6	70	75	80	4
30. Foot length	230	255	280	15	215	235	255	12
31. Foot breadth	85	95	105	7	80	90	100	6
32. Span	1480	1670	1860	114	1450	1580	1710	79
33. Elbow span	775	880	985	65	755	835	915	48
34. Vertical grip reach (standing)	1825	1990	2155	101	1765	1930	2095	101
35. Vertical grip reach (sitting)	1015	1140	1270	77	980	1105	1235	77
36. Forward grip reach	615	680	745	41	595	655	715	36

See notes on p. 82.

Table 4.20. Anthropometric estimates for British 15-year-olds (all dimensions in millimetres).

Dimension	Boys				Girls			
	5th %ile	50th %ile	95th %ile	SD	5th %ile	50th %ile	95th %ile	SD
1. Stature	1555	1690	1825	83	1510	1610	1710	62
2. Eye height	1430	1570	1710	84	1395	1495	1595	62
3. Shoulder height	1265	1385	1505	73	1215	1310	1405	58
4. Elbow height	965	1055	1145	56	915	995	1075	48
5. Hip height	825	895	965	44	745	815	885	42
6. Knuckle height	650	725	800	45	650	715	780	39
7. Fingertip height	530	605	680	45	540	605	670	39
8. Sitting height	785	870	955	51	790	845	900	33
9. Sitting eye height	680	755	830	47	680	735	790	33
10. Sitting shoulder height	495	555	615	36	490	535	580	27
11. Sitting elbow height	170	225	280	34	180	225	270	28
12. Thigh thickness	115	140	165	15	115	140	165	14
13. Buttock–knee length	515	570	625	32	505	550	595	27
14. Buttock–popliteal length	425	480	535	32	435	470	505	22
15. Knee height	485	535	585	31	450	495	540	26
16. Popliteal height	385	430	475	28	360	400	440	25
17. Shoulder breadth (bideltoid)	354	415	465	30	350	390	430	23
18. Shoulder breadth (biacromial)	330	370	410	23	320	350	380	18
19. Hip breadth	275	320	365	26	295	335	375	25
20. Chest (bust) depth	155	205	255	30	175	215	260	26
21. Abdominal depth	180	220	260	24	185	220	255	22
22. Shoulder–elbow length	325	355	385	19	305	335	365	17
23. Elbow–fingertip length	420	460	500	25	395	425	455	17
24. Upper limb length	695	770	845	47	650	705	760	33
25. Shoulder–grip length	570	650	725	47	540	595	650	33
26. Head length	185	195	205	7	170	180	190	7
27. Head breadth	145	155	165	6	130	140	150	5
28. Hand length	170	185	200	10	155	170	185	9
29. Hand breadth	75	85	95	5	70	75	80	4
30. Foot length	240	260	280	13	215	235	255	13
31. Foot breadth	85	95	105	7	80	90	100	5
32. Span	1560	1740	1920	109	1490	1600	1710	67
33. Elbow span	815	915	1020	62	780	845	910	41
34. Vertical grip reach (standing)	1900	2060	2220	97	1810	1960	2110	91
35. Vertical grip reach (sitting)	1075	1190	1310	71	1005	1120	1240	71
36. Forward grip reach	635	700	765	40	600	665	730	39

See notes on p. 82.

Table 4.21. Anthropometric estimates for British 16-year-olds (all dimensions in millimetres).

	Men				Women			
Dimension	5th %ile	50th %ile	95th %ile	SD	5th %ile	50th %ile	95th %ile	SD
1. Stature	1620	1730	1840	68	1520	1620	1720	61
2. Eye height	1500	1610	1720	67	1410	1510	1610	60
3. Shoulder height	1315	1415	1515	62	1225	1315	1405	55
4. Elbow height	995	1075	1155	49	930	1005	1080	45
5. Hip height	830	910	990	49	755	820	885	40
6. Knuckle height	675	740	805	40	660	720	780	36
7. Fingertip height	555	620	685	40	545	605	665	36
8. Sitting height	830	895	960	39	800	855	910	33
9. Sitting eye height	725	785	845	35	685	740	795	32
10. Sitting shoulder height	520	570	620	29	500	545	590	27
11. Sitting elbow height	190	235	280	28	185	230	275	26
12. Thigh thickness	125	150	175	15	120	145	170	14
13. Buttock–knee length	530	580	630	29	510	555	600	27
14. Buttock–popliteal length	435	490	545	32	435	480	525	26
15. Knee height	500	545	590	27	450	495	540	26
16. Popliteal height	395	440	485	28	365	405	445	25
17. Shoulder breadth (bideltoid)	380	430	480	29	360	395	430	21
18. Shoulder breadth (biacromial)	340	380	420	23	330	355	380	16
19. Hip breadth	290	330	370	23	305	345	385	25
20. Chest (bust) depth	165	215	265	29	180	225	265	25
21. Abdominal depth	185	225	265	24	185	220	255	21
22. Shoulder–elbow length	335	365	395	18	310	335	365	17
23. Elbow–fingertip length	435	470	505	22	395	425	455	17
24. Upper limb length	725	790	855	40	660	710	760	29
25. Shoulder–grip length	605	670	735	40	550	595	645	29
26. Head length	185	195	205	7	165	180	195	8
27. Head breadth	145	155	165	5	135	145	155	5
28. Hand length	170	185	200	9	160	175	190	9
29. Hand breadth	80	85	90	4	70	75	80	4
30. Foot length	240	260	280	12	220	240	260	12
31. Foot breadth	90	100	110	6	80	90	100	5
32. Span	1640	1785	1930	88	1500	1610	1720	67
33. Elbow span	860	940	1025	50	785	850	920	41
34. Vertical grip reach (standing)	1945	2100	2255	93	1820	1965	2110	88
35. Vertical grip reach (sitting)	1130	1225	1320	58	1035	1135	1230	58
36. Forward grip reach	650	720	790	42	605	670	735	41

See notes on p. 82.

Table 4.22. Anthropometric estimates for British 17-year-olds (all dimensions in millimetres).

	Boys				Girls			
Dimension	5th %ile	50th %ile	95th %ile	SD	5th %ile	50th %ile	95th %ile	SD
1. Stature	1640	1750	1860	66	1520	1620	1720	61
2. Eye height	1530	1635	1740	65	1420	1515	1610	58
3. Shoulder height	1335	1435	1535	62	1235	1320	1405	52
4. Elbow height	1010	1090	1170	50	935	1005	1075	43
5. Hip height	845	925	1005	50	755	820	885	39
6. Knuckle height	690	755	820	41	670	725	780	33
7. Fingertip height	565	630	700	41	555	610	665	33
8. Sitting height	850	910	970	35	800	855	910	33
9. Sitting eye height	745	795	845	31	690	740	790	30
10. Sitting shoulder height	535	585	635	29	515	555	595	25
11. Sitting elbow height	195	240	285	28	190	230	270	25
12. Thigh thickness	125	155	185	17	120	145	170	16
13. Buttock–knee length	535	585	635	30	515	560	605	28
14. Buttock–popliteal length	445	495	545	30	435	480	525	27
15. Knee height	505	550	595	27	455	500	545	26
16. Popliteal height	405	445	485	25	365	405	445	25
17. Shoulder breadth (bideltoid)	400	445	490	28	360	395	430	21
18. Shoulder breadth (biacromial)	350	385	420	21	335	360	385	16
19. Hip breadth	295	335	375	24	300	345	390	28
20. Chest (bust) depth	180	225	270	27	190	230	270	24
21. Abdominal depth	195	235	275	25	185	220	255	21
22. Shoulder–elbow length	335	365	395	19	305	335	365	18
23. Elbow–fingertip length	440	475	510	21	395	425	455	17
24. Upper limb length	730	790	850	36	660	710	760	29
25. Shoulder–grip length	605	665	725	36	550	595	645	29
26. Head length	185	200	215	8	165	180	195	8
27. Head breadth	145	155	165	6	135	145	155	5
28. Hand length	175	190	205	9	160	175	190	9
29. Hand breadth	80	90	100	5	70	75	80	4
30. Foot length	240	265	290	15	220	240	260	12
31. Foot breadth	90	100	110	5	80	90	100	5
32. Span	1660	1795	1930	81	1510	1615	1720	64
33. Elbow span	870	945	1020	46	790	855	915	39
34. Vertical grip reach (standing)	1980	2125	2270	87	1830	1970	2110	85
35. Vertical grip reach (sitting)	1145	1240	1330	57	1050	1145	1235	57
36. Forward grip reach	655	730	805	46	610	670	730	37

See notes on p. 82.

Table 4.23. Anthropometric estimates for British 18-year-olds (all dimensions in millimetres).

	Men				Women			
Dimension	5th %ile	50th %ile	95th %ile	SD	5th %ile	50th %ile	95th %ile	SD
1. Stature	1660	1760	1860	60	1530	1620	1710	56
2. Eye height	1555	1650	1745	59	1430	1520	1610	55
3. Shoulder height	1355	1445	1535	54	1235	1320	1405	52
4. Elbow height	1010	1105	1175	44	940	1010	1080	42
5. Hip height	865	935	1005	43	755	820	885	40
6. Knuckle height	705	765	825	35	670	725	780	32
7. Fingertip height	585	640	700	35	560	610	665	32
8. Sitting height	860	915	970	32	800	855	910	32
9. Sitting eye height	745	800	855	32	695	745	795	30
10. Sitting shoulder height	550	600	650	30	515	560	605	28
11. Sitting elbow height	200	245	290	26	185	230	275	26
12. Thigh thickness	135	160	185	15	120	145	170	14
13. Buttock–knee length	545	590	635	26	515	560	605	28
14. Buttock–popliteal length	450	500	550	29	435	480	525	27
15. Knee height	505	550	595	26	455	500	545	26
16. Popliteal height	405	445	485	25	365	405	445	25
17. Shoulder breadth (bideltoid)	415	455	495	23	360	395	430	21
18. Shoulder breadth (biacromial)	365	395	425	17	335	360	385	16
19. Hip breadth	300	340	380	25	300	345	390	27
20. Chest (bust) depth	190	225	260	21	195	235	275	24
21. Abdominal depth	205	240	275	21	185	220	255	20
22. Shoulder–elbow length	340	370	395	17	310	335	360	16
23. Elbow–fingertip length	450	480	510	18	395	425	455	17
24. Upper limb length	740	790	840	31	660	710	760	29
25. Shoulder–grip length	615	665	715	31	550	595	645	29
26. Head length	185	200	215	8	170	180	190	7
27. Head breadth	145	155	165	5	135	145	155	5
28. Hand length	175	190	205	8	160	175	190	8
29. Hand breadth	85	90	95	4	70	75	80	4
30. Foot length	250	270	290	12	220	240	260	11
31. Foot breadth	90	100	110	5	80	90	100	5
32. Span	1695	1810	1925	71	1520	1620	1720	62
33. Elbow span	890	955	1020	40	795	855	920	38
34. Vertical grip reach (standing)	2045	2150	2255	65	1830	1970	2110	85
35. Vertical grip reach (sitting)	1170	1250	1335	52	1065	1150	1235	52
36. Forward grip reach	675	740	805	41	610	670	730	37

See notes on p. 82.

Table 4.24. Anthropometric estimates for British adults aged 19–25 years (all dimensions in millimetres).

Dimension	Men 5th %ile	Men 50th %ile	Men 95th %ile	Men SD	Women 5th %ile	Women 50th %ile	Women 95th %ile	Women SD
1. Stature	1640	1760	1880	73	1520	1620	1720	61
2. Eye height	1530	1650	1770	72	1415	1515	1615	60
3. Shoulder height	1330	1445	1555	69	1225	1320	1410	57
4. Elbow height	1020	1105	1195	54	940	1015	1090	45
5. Hip height	850	935	1020	52	745	815	885	43
6. Knuckle height	695	765	835	42	665	725	785	35
7. Fingertip height	595	665	730	40	565	630	690	38
8. Sitting height	855	915	980	37	800	855	915	35
9. Sitting eye height	740	795	855	36	690	745	800	33
10. Sitting shoulder height	545	600	655	33	510	560	610	31
11. Sitting elbow height	195	245	300	32	180	230	275	28
12. Thigh thickness	130	160	185	16	120	150	175	16
13. Buttock–knee length	545	595	650	32	520	565	615	29
14. Buttock–popliteal length	445	500	555	34	430	475	525	29
15. Knee height	495	550	605	33	460	500	545	26
16. Popliteal height	400	445	495	30	355	400	445	27
17. Shoulder breadth (bideltoid)	415	465	510	29	355	395	435	24
18. Shoulder breadth (biacromial)	370	405	440	21	330	360	390	18
19. Hip breadth	300	350	400	31	300	350	400	29
20. Chest (bust) depth	185	225	270	26	190	235	275	26
21. Abdominal depth	195	240	280	26	185	220	260	22
22. Shoulder–elbow length	335	370	405	21	305	330	360	17
23. Elbow–fingertip length	445	480	515	22	400	430	465	19
24. Upper limb length	730	790	850	37	660	710	760	32
25. Shoulder–grip length	615	670	730	34	560	605	650	29
26. Head length	185	195	210	8	170	180	190	7
27. Head breadth	145	155	165	7	135	145	155	5
28. Hand length	175	190	210	10	160	175	190	9
29. Hand breadth	80	90	95	5	70	75	85	4
30. Foot length	245	270	290	15	220	240	260	12
31. Foot breadth	90	100	110	7	80	90	100	5
32. Span	1670	1815	1955	86	1500	1615	1730	70
33. Elbow span	875	955	1035	49	785	855	925	42
34. Vertical grip reach (standing)	1950	2085	2220	83	1805	1915	2030	70
35. Vertical grip reach (sitting)	1155	1260	1360	63	1070	1155	1245	52
36. Forward grip reach	730	790	845	36	655	705	755	31

Note: Best estimate for overall British population in the year 2000.
See also notes on p. 82.

Table 4.25. Anthropometric estimates for British adults aged 19–45 years (all dimensions in millimetres).

Dimension	Men 5th %ile	Men 50th %ile	Men 95th %ile	Men SD	Women 5th %ile	Women 50th %ile	Women 95th %ile	Women SD
1. Stature	1635	1745	1860	69	1515	1615	1715	61
2. Eye height	1525	1635	1750	68	1415	1515	1615	60
3. Shoulder height	1325	1435	1540	65	1225	1315	1410	57
4. Elbow height	1015	1100	1185	51	940	1015	1085	45
5. Hip height	845	925	1005	49	745	815	885	43
6. Knuckle height	695	760	825	40	665	725	780	35
7. Fingertip height	595	660	720	38	565	625	690	38
8. Sitting height	855	915	970	35	800	855	915	35
9. Sitting eye height	740	795	850	34	690	745	800	33
10. Sitting shoulder height	545	595	650	31	510	560	610	31
11. Sitting elbow height	195	245	295	30	190	235	280	28
12. Thigh thickness	135	160	185	15	125	155	180	16
13. Buttock–knee length	545	595	645	30	520	570	610	29
14. Buttock–popliteal length	445	495	550	32	435	480	530	29
15. Knee height	495	545	600	31	460	500	545	26
16. Popliteal height	395	445	490	28	355	400	445	27
17. Shoulder breadth (bideltoid)	420	465	510	28	355	395	435	24
18. Shoulder breadth (biacromial)	365	400	435	20	330	360	390	18
19. Hip breadth	310	355	405	29	300	365	425	37
20. Chest (bust) depth	200	240	275	23	195	240	285	26
21. Abdominal depth	210	255	300	28	195	245	290	29
22. Shoulder–elbow length	330	365	400	20	305	330	360	17
23. Elbow–fingertip length	445	475	510	21	400	430	460	19
24. Upper limb length	725	785	840	35	655	710	760	32
25. Shoulder–grip length	615	665	720	32	555	605	650	29
26. Head length	185	195	210	8	165	180	190	7
27. Head breadth	145	155	165	6	135	145	155	5
28. Hand length	175	190	205	10	160	175	190	9
29. Hand breadth	80	85	95	5	70	75	85	4
30. Foot length	245	265	290	14	215	235	255	12
31. Foot breadth	90	100	110	6	80	90	100	5
32. Span	1665	1800	1935	81	1500	1615	1730	70
33. Elbow span	875	950	1025	46	785	855	920	42
34. Vertical grip reach (standing)	1940	2070	2200	79	1800	1915	2030	70
35. Vertical grip reach (sitting)	1150	1250	1345	59	1070	1155	1240	52
36. Forward grip reach	725	785	840	34	655	705	755	31

See notes on p. 82.

Table 4.26. Anthropometric estimates for British adults aged 45–65 years (all dimensions in millimetres).

Dimension	Men 5th %ile	Men 50th %ile	Men 95th %ile	Men SD	Women 5th %ile	Women 50th %ile	Women 95th %ile	Women SD
1. Stature	1610	1720	1830	67	1495	1595	1695	61
2. Eye height	1505	1610	1720	66	1395	1495	1595	60
3. Shoulder height	1305	1410	1515	63	1205	1300	1395	57
4. Elbow height	1000	1080	1160	50	925	1000	1075	45
5. Hip height	835	910	990	48	735	805	875	43
6. Knuckle height	685	750	810	39	655	715	775	35
7. Fingertip height	590	650	710	37	555	620	680	38
8. Sitting height	840	900	955	34	785	845	900	35
9. Sitting eye height	725	780	835	34	680	735	790	33
10. Sitting shoulder height	535	585	635	30	500	550	600	31
11. Sitting elbow height	190	240	290	29	185	230	280	28
12. Thigh thickness	135	160	185	15	130	155	180	16
13. Buttock–knee length	540	585	635	29	520	570	620	29
14. Buttock–popliteal length	440	490	540	31	435	480	530	29
15. Knee height	490	540	590	30	450	495	540	26
16. Popliteal height	390	435	480	27	350	395	440	27
17. Shoulder breadth (bideltoid)	415	460	505	27	350	390	430	24
18. Shoulder breadth (biacromial)	360	395	425	19	325	355	385	18
19. Hip breadth	310	360	405	28	315	375	440	37
20. Chest (bust) depth	225	260	295	20	220	265	305	26
21. Abdominal depth	230	285	340	34	220	270	320	31
22. Shoulder–elbow length	330	360	390	19	300	325	355	17
23. Elbow–fingertip length	435	470	505	20	395	425	455	19
24. Upper limb length	715	770	830	34	650	700	755	32
25. Shoulder–grip length	605	655	710	31	550	595	645	29
26. Head length	180	195	205	7	165	175	190	7
27. Head breadth	140	150	160	6	135	140	150	5
28. Hand length	170	185	205	9	155	170	185	9
29. Hand breadth	80	85	95	5	70	75	80	4
30. Foot length	240	260	285	13	215	235	255	12
31. Foot breadth	85	95	105	6	80	90	95	5
32. Span	1640	1770	1900	79	1480	1595	1710	70
33. Elbow span	860	935	1010	45	775	845	910	42
34. Vertical grip reach (standing)	1910	2035	2160	76	1775	1890	2000	70
35. Vertical grip reach (sitting)	1135	1230	1325	58	1055	1140	1225	52
36. Forward grip reach	715	770	825	33	645	695	750	31

See notes on p. 82.

Table 4.27. Anthropometric estimates for British adults aged 65–80 years (all dimensions in millimetres).

Dimension	Men 5th %ile	Men 50th %ile	Men 95th %ile	Men SD	Women 5th %ile	Women 50th %ile	Women 95th %ile	Women SD
1. Stature	1575	1685	1790	66	1475	1570	1670	60
2. Eye height	1470	1575	1685	65	1375	1475	1570	59
3. Shoulder height	1280	1380	1480	62	1190	1280	1375	56
4. Elbow height	975	1055	1135	49	910	985	1055	44
5. Hip height	820	895	975	47	740	810	875	42
6. Knuckle height	670	730	795	38	645	705	760	35
7. Fingertip height	575	635	695	36	550	610	670	37
8. Sitting height	815	875	930	36	750	815	885	41
9. Sitting eye height	705	760	815	34	645	710	770	38
10. Sitting shoulder height	520	570	625	32	475	535	590	36
11. Sitting elbow height	175	220	270	29	165	210	260	28
12. Thigh thickness	125	150	175	15	115	145	170	16
13. Buttock–knee length	530	580	625	29	520	565	615	29
14. Buttock–popliteal length	430	485	535	31	430	480	525	29
15. Knee height	480	525	575	30	455	500	540	26
16. Popliteal height	385	425	470	27	355	395	440	26
17. Shoulder breadth (bideltoid)	400	445	485	26	345	385	425	23
18. Shoulder breadth (biacromial)	350	375	405	17	320	350	380	17
19. Hip breadth	305	350	395	28	310	370	430	37
20. Chest (bust) depth	225	260	290	20	220	265	305	26
21. Abdominal depth	245	300	355	33	225	270	320	30
22. Shoulder–elbow length	320	350	385	19	295	320	350	17
23. Elbow–fingertip length	425	460	490	20	390	420	450	19
24. Upper limb length	700	755	810	34	640	690	740	31
25. Shoulder–grip length	595	645	695	30	540	590	635	28
26. Head length	175	190	200	7	165	175	185	7
27. Head breadth	140	150	160	6	130	140	150	5
28. Hand length	170	185	200	9	155	170	185	9
29. Hand breadth	75	85	90	5	65	75	80	4
30. Foot length	235	255	280	13	210	230	250	12
31. Foot breadth	85	95	105	6	80	85	95	5
32. Span	1605	1735	1860	78	1460	1570	1685	68
33. Elbow span	840	915	985	44	760	830	900	41
34. Vertical grip reach (standing)	1840	1965	2090	75	1725	1835	1950	68
35. Vertical grip reach (sitting)	1110	1205	1295	57	1040	1125	1210	52
36. Forward grip reach	700	755	805	32	640	685	735	30

See notes on p. 83.

Table 4.28. Anthropometric estimates for 'elderly people' (all dimensions in millimetres).

Dimension	Men				Women			
	5th %ile	50th %ile	95th %ile	SD	5th %ile	50th %ile	95th %ile	SD
1. Stature	1515	1640	1765	77	1400	1515	1630	70
2. Eye height	1410	1535	1660	76	1305	1420	1535	69
3. Shoulder height	1225	1345	1465	72	1130	1235	1340	65
4. Elbow height	935	1025	1120	57	860	945	1030	52
5. Hip height	785	875	965	55	700	780	860	49
6. Knuckle height	640	715	785	45	610	680	745	41
7. Fingertip height	550	620	690	42	515	590	660	43
8. Sitting height	785	850	920	42	710	785	865	48
9. Sitting eye height	675	740	805	40	610	685	755	45
10. Sitting shoulder height	495	555	615	37	445	515	585	42
11. Sitting elbow height	160	215	270	34	150	205	255	32
12. Thigh thickness	120	145	175	17	105	140	170	19
13. Buttock–knee length	510	565	620	34	490	545	600	34
14. Buttock–popliteal length	410	470	530	36	405	460	515	34
15. Knee height	455	515	570	35	430	480	530	30
16. Popliteal height	365	415	470	32	330	380	430	31
17. Shoulder breadth (bideltoid)	380	430	480	31	325	370	415	27
18. Shoulder breadth (biacromial)	335	365	400	20	305	335	370	20
19. Hip breadth	290	340	395	32	285	355	425	43
20. Chest (bust) depth	215	255	290	23	205	255	305	30
21. Abdominal depth	230	290	355	39	205	260	320	35
22. Shoulder–elbow length	305	345	380	22	280	310	345	20
23. Elbow–fingertip length	410	450	485	23	370	405	440	22
24. Upper limb length	670	735	800	39	605	665	725	36
25. Shoulder–grip length	570	625	685	35	510	565	620	33
26. Head length	170	185	200	8	155	170	180	8
27. Head breadth	135	145	155	7	125	135	145	6
28. Hand length	160	180	195	11	145	165	180	10
29. Hand breadth	75	80	90	5	65	70	80	5
30. Foot length	225	250	275	15	200	225	245	14
31. Foot breadth	80	90	105	7	75	85	95	6
32. Span	1540	1690	1840	91	1380	1515	1645	80
33. Elbow span	805	890	975	52	720	800	880	48
34. Vertical grip reach (standing)	1770	1915	2060	88	1640	1770	1900	80
35. Vertical grip reach (sitting)	1065	1175	1280	66	985	1085	1180	60
36. Forward grip reach	675	735	795	38	605	660	720	35

See notes on p. 83.

Table 4.29. Anthropometric estimates for US adults (all dimensions in millimetres).

Dimension	Men 5th %ile	Men 50th %ile	Men 95th %ile	Men SD	Women 5th %ile	Women 50th %ile	Women 95th %ile	Women SD
1. Stature	1640	1755	1870	71	1520	1625	1730	64
2. Eye height	1595	1710	1825	70	1420	1525	1630	63
3. Shoulder height	1330	1440	1550	67	1225	1325	1425	60
4. Elbow height	1020	1105	1190	53	945	1020	1095	47
5. Hip height	835	915	995	50	760	835	910	45
6. Knuckle height	700	765	830	41	670	730	790	37
7. Fingertip height	595	660	725	39	565	630	695	40
8. Sitting height	855	915	975	36	800	860	920	36
9. Sitting eye height	740	800	860	35	690	750	810	35
10. Sitting shoulder height	545	600	655	32	510	565	620	32
11. Sitting elbow height	195	245	295	31	185	235	285	29
12. Thigh thickness	135	160	185	16	125	155	185	17
13. Buttock–knee length	550	600	650	31	525	575	625	31
14. Buttock–popliteal length	445	500	555	33	440	490	540	31
15. Knee height	495	550	605	32	460	505	550	28
16. Popliteal height	395	445	495	29	360	405	450	28
17. Shoulder breadth (bideltoid)	425	470	515	28	360	400	440	25
18. Shoulder breadth (biacromial)	365	400	435	21	330	360	390	19
19. Hip breadth	310	360	410	30	310	375	440	39
20. Chest (bust) depth	220	255	290	22	210	255	300	28
21. Abdominal depth	220	275	330	32	210	260	310	31
22. Shoulder–elbow length	330	365	400	21	305	335	365	18
23. Elbow–fingertip length	445	480	515	21	400	435	470	20
24. Upper limb length	730	790	850	36	655	715	775	35
25. Shoulder–grip length	615	670	725	33	560	610	660	30
26. Head length	180	195	210	8	165	180	195	8
27. Head breadth	145	155	165	6	135	145	155	6
28. Hand length	175	191	205	10	160	175	190	10
29. Hand breadth	80	90	100	5	65	75	85	5
30. Foot length	240	265	290	14	220	240	260	13
31. Foot breadth	90	100	110	6	80	90	100	6
32. Span	1670	1810	1950	84	1505	1625	1745	73
33. Elbow span	875	955	1035	48	790	860	930	44
34. Vertical grip reach (standing)	1950	2080	2210	80	1805	1925	2045	73
35. Vertical grip reach (sitting)	1155	1255	1355	61	1070	1160	1250	55
36. Forward grip reach	725	785	845	35	655	710	765	32

See notes on p. 83.

Table 4.30. Anthropometric estimates for French adults (all dimensions in millimetres).

Dimension	Men 5th %ile	Men 50th %ile	Men 95th %ile	Men SD	Women 5th %ile	Women 50th %ile	Women 95th %ile	Women SD
1. Stature	1600	1715	1830	69	1500	1600	1700	61*
2. Eye height	1450	1560	1670	68	1400	1500	1600	60
3. Shoulder height	1300	1405	1510	65	1210	1305	1400	57
4. Elbow height	995	1080	1165	51	925	1000	1075	45
5. Hip height	815	895	975	49	750	820	890	43
6. Knuckle height	680	745	810	40	655	715	775	35
7. Fingertip height	580	645	710	38	560	620	680	37
8. Sitting height	850	910	970	35	810	860	910	31*
9. Sitting eye height	735	795	855	35	700	750	800	30*
10. Sitting shoulder height	570	620	670	31	535	580	625	27*
11. Sitting elbow height	190	240	290	30	185	230	275	28
12. Thigh thickness	150	180	210	17	135	165	195	17*
13. Buttock–knee length	550	595	640	28	520	565	610	28*
14. Buttock–popliteal length	435	480	525	26	415	460	505	26*
15. Knee height	485	530	575	26	455	495	535	24*
16. Popliteal height	385	425	465	25	350	390	430	23*
17. Shoulder breadth (bideltoid)	425	470	515	26	380	425	470	27*
18. Shoulder breadth (biacromial)	360	395	430	20	325	355	385	18
19. Hip breadth	330	370	410	24	330	380	430	30*
20. Chest (bust) depth	210	245	280	22	205	250	295	26
21. Abdominal depth	220	270	320	31	205	255	305	30
22. Shoulder–elbow length	325	360	395	20	300	330	360	17*
23. Elbow–fingertip length	435	470	505	21	395	425	455	19*
24. Upper limb length	710	770	830	35	650	705	760	32*
25. Shoulder–grip length	600	655	710	32	550	600	650	29
26. Head length	175	190	205	8	170	180	190	7
27. Head breadth	140	150	160	6	130	140	150	5
28. Hand length	170	185	200	10	160	175	190	9
29. Hand breadth	75	85	95	5	70	75	80	4
30. Foot length	235	260	285	14	215	235	255	12
31. Foot breadth	85	95	105	6	80	90	100	5
32. Span	1630	1765	1900	81	1485	1600	1715	70
33. Elbow span	855	930	1005	46	775	845	915	42
34. Vertical grip reach (standing)	1900	2030	2160	79	1780	1895	2010	70
35. Vertical grip reach (sitting)	1130	1225	1320	59	1060	1145	1230	52
36. Forward grip reach	715	770	825	34	645	700	755	32

See notes on p. 83.

Table 4.31. Anthropometric estimates for German adults (all dimensions in millimetres).

Dimension	Men				Women			
	5th %ile	50th %ile	95th %ile	SD	5th %ile	50th %ile	95th %ile	SD
1. Stature	1645	1745	1845	62	1520	1635	1750	69*
2. Eye height	1535	1635	1735	61	1420	1530	1640	68
3. Shoulder height	1370	1465	1560	58	1240	1320	1400	50*
4. Elbow height	1020	1095	1170	46	925	1000	1075	46*(W)
5. Hip height	840	910	980	44	760	840	920	48
6. Knuckle height	700	760	820	36	665	730	795	40
7. Fingertip height	605	660	715	34	565	635	705	43
8. Sitting height	865	920	975	32	800	865	930	39*(M)
9. Sitting eye height	750	800	850	31	680	740	800	37*
10. Sitting shoulder height	550	595	640	28	480	525	570	27*(W)
11. Sitting elbow height	195	235	275	25	165	205	245	23*
12. Thigh thickness	35	150	265	70	125	155	185	19*(M)
13. Buttock–knee length	560	600	640	25	525	580	635	33*(M)
14. Buttock–popliteal length	445	495	545	29	435	490	545	33
15. Knee height	500	545	590	28	455	505	555	30
16. Popliteal height	415	455	495	25	355	395	435	23*
17. Shoulder breadth (bideltoid)	425	465	505	23	355	400	445	27
18. Shoulder breadth (biacromial)	370	400	430	17	325	360	395	20*(M)
19. Hip breadth	315	350	385	21	305	375	445	42*(M)
20. Chest (bust) depth	215	250	285	20	205	255	305	30
21. Abdominal depth	230	275	320	28	205	260	315	33
22. Shoulder–elbow length	335	365	395	18	305	335	365	19
23. Elbow–fingertip length	445	475	505	19	400	435	470	21
24. Upper limb length	735	785	835	31	660	720	780	36
25. Shoulder–grip length	615	665	715	29	555	610	665	32
26. Head length	185	195	205	7	165	180	195	8*(M)
27. Head breadth	145	155	165	5	135	145	155	6*(M)
28. Hand length	170	185	200	10	160	175	190	10*(M)
29. Hand breadth	80	85	90	4	65	75	85	5*(M)
30. Foot length	240	260	280	12	215	240	265	14*(M)
31. Foot breadth	90	100	110	6	80	90	100	6
32. Span	1675	1795	1915	73	1505	1635	1765	79
33. Elbow span	880	950	1020	42	785	865	945	48
34. Vertical grip reach (standing)	1950	2065	2180	71	1805	1935	2065	79
35. Vertical grip reach (sitting)	1160	1245	1330	53	1075	1170	1265	59
36. Forward grip reach	730	780	830	30	655	715	775	35

See notes on p. 83.

Table 4.32. Anthropometric estimates for Swedish adults (all dimensions in millimetres).

	Men				Women			
Dimension	5th %ile	50th %ile	95th %ile	SD	5th %ile	50th %ile	95th %ile	SD
1. Stature	1630	1740	1850	68	1540	1640	1740	62*
2. Eye height	1520	1630	1740	68	1435	1535	1635	62*
3. Shoulder height	1345	1445	1545	62	1255	1355	1455	60*
4. Elbow height	1020	1100	1180	49	905	1025	1145	73*
5. Hip height	815	890	965	45	745	830	915	52*
6. Knuckle height	720	760	800	25	675	735	795	36
7. Fingertip height	595	655	715	37	570	635	700	38
8. Sitting height	830	900	970	43	805	860	915	33*
9. Sitting eye height	715	785	855	42	705	755	805	30*
10. Sitting shoulder height	545	600	655	34	525	575	625	30*
11. Sitting elbow height	175	225	275	31	165	215	265	31*
12. Thigh thickness	120	152	180	18	130	155	180	16*
13. Buttock–knee length	545	595	645	30	525	585	645	35*
14. Buttock–popliteal length	430	480	530	30	430	485	540	33*
15. Knee height	480	530	580	30	455	500	545	28*
16. Popliteal height	385	430	475	27	350	400	450	29
17. Shoulder breadth (bideltoid)	420	465	510	27	355	390	425	20
18. Shoulder breadth (biacromial)	365	400	435	20	325	350	375	15*(W)
19. Hip breadth	310	360	410	29	690	755	820	38
20. Chest (bust) depth	185	220	255	21	185	241	300	35
21. Abdominal depth	190	240	290	31	180	245	310	40*
22. Shoulder–elbow length	330	365	400	20	305	335	365	17
23. Elbow–fingertip length	440	475	510	20	160	175	190	10*(W)
24. Upper limb length	720	780	840	35	660	705	750	28*(W)
25. Shoulder–grip length	615	665	715	31	555	595	635	24
26. Head length	185	195	205	7	170	180	190	7
27. Head breadth	145	155	165	6	135	145	155	6
28. Hand length	175	190	205	10	165	180	195	10*(W)
29. Hand breadth	75	85	95	5	70	75	80	4*(W)
30. Foot length	240	265	290	14	225	245	265	11*(W)
31. Foot breadth	85	95	105	6	85	95	105	7*(W)
32. Span	1660	1790	1920	80	1525	1640	1755	71
33. Elbow span	870	945	1020	45	795	865	935	43
34. Vertical grip reach (standing)	1930	2060	2190	78	1825	1940	2055	70
35. Vertical grip reach (sitting)	1150	1245	1340	58	1090	1175	1260	53
36. Forward grip reach	725	780	835	33	665	715	765	31

See notes on p. 83.

Table 4.33. Anthropometric estimates for Swiss adults (all dimensions in millimetres).

Dimension	Men 5th %ile	Men 50th %ile	Men 95th %ile	Men SD	Women 5th %ile	Women 50th %ile	Women 95th %ile	Women SD
1. Stature	1535	1690	1845	95	1415	1590	1765	105*
2. Eye height	1430	1585	1740	95	1320	1490	1660	104
3. Shoulder height	1255	1410	1565	94	1125	1320	1515	118*
4. Elbow height	945	1060	1175	70	865	995	1125	78
5. Hip height	770	880	990	68	695	815	935	74
6. Knuckle height	645	735	825	55	610	712	810	61
7. Fingertip height	550	635	720	52	510	615	720	65
8. Sitting height	805	885	965	48	740	840	940	60
9. Sitting eye height	695	770	845	47	635	730	825	57
10. Sitting shoulder height	505	575	645	43	465	550	635	53
11. Sitting elbow height	165	235	305	42	150	230	310	48
12. Thigh thickness	120	155	190	21	110	155	200	28
13. Buttock–knee length	510	560	610	31	500	550	600	30
14. Buttock–popliteal length	415	470	525	33	415	465	515	30*
15. Knee height	475	520	565	27	430	470	510	25*
16. Popliteal height	420	455	490	22	340	375	410	22*
17. Shoulder breadth (bideltoid)	395	435	475	25	365	410	455	27*
18. Shoulder breadth (biacromial)	340	385	430	28	300	350	400	30
19. Hip breadth	285	350	415	40	260	365	470	64
20. Chest (bust) depth	195	245	295	30	175	250	325	45
21. Abdominal depth	195	265	335	43	170	255	340	51
22. Shoulder–elbow length	320	365	410	28	300	340	380	25*
23. Elbow–fingertip length	440	475	510	22	400	440	480	25*
24. Upper limb length	680	760	840	48	610	700	790	55
25. Shoulder–grip length	575	645	715	44	515	595	675	49
26. Head length	175	190	205	10	155	175	195	13
27. Head breadth	135	150	165	9	125	140	155	9
28. Hand length	165	185	205	13	145	170	195	16
29. Hand breadth	75	85	95	7	65	75	85	7
30. Foot length	225	255	285	19	200	235	270	21
31. Foot breadth	80	95	110	9	75	90	105	9
32. Span	1575	1740	1905	100	1430	1590	1750	98*
33. Elbow span	815	920	1025	64	720	840	960	72
34. Vertical grip reach (standing)	1820	2000	2180	108	1685	1880	2075	120
35. Vertical grip reach (sitting)	1075	1210	1345	82	985	1135	1285	90
36. Forward grip reach	685	760	835	47	610	695	780	53

See notes on p. 83.

Table 4.34. Anthropometric estimates for Polish adults (all dimensions in millimetres).

Dimension	Men 5th %ile	Men 50th %ile	Men 95th %ile	Men SD	Women 5th %ile	Women 50th %ile	Women 95th %ile	Women SD
1. Stature	1595	1695	1795	61	1480	1575	1670	58*
2. Eye height	1505	1600	1695	58	1390	1485	1580	57*
3. Shoulder height	1275	1365	1455	54	1170	1280	1390	68*
4. Elbow height	990	1065	1140	45	915	985	1055	43
5. Hip height	810	880	950	43	745	810	875	41
6. Knuckle height	545	595	640	29	535	570	605	22*
7. Fingertip height	615	675	735	36	590	645	700	32*
8. Sitting height	830	885	940	34	770	825	880	33*
9. Sitting eye height	720	780	840	36	665	725	785	35*
10. Sitting shoulder height	555	605	655	31	515	565	615	31*
11. Sitting elbow height	195	240	285	27	185	230	275	27
12. Thigh thickness	110	140	170	19	115	140	165	16*
13. Buttock–knee length	540	585	630	26	515	565	615	29*
14. Buttock–popliteal length	405	455	505	29	360	450	540	54*
15. Knee height	485	530	575	27	445	485	525	24*
16. Popliteal height	410	445	480	21	390	420	450	19*
17. Shoulder breadth (bideltoid)	405	440	475	21	350	380	410	18*
18. Shoulder breadth (biacromial)	360	390	420	18	320	350	380	17*
19. Hip breadth	305	345	385	25	315	360	405	26*
20. Chest (bust) depth	215	245	275	19	205	245	285	25
21. Abdominal depth	220	265	310	27	205	250	295	28
22. Shoulder–elbow length	310	330	350	13	280	300	320	12*
23. Elbow–fingertip length	430	460	490	18	390	420	450	18
24. Upper limb length	705	755	805	30	655	700	745	28*
25. Shoulder–grip length	640	675	710	22	595	625	655	18*
26. Head length	175	185	195	7	170	180	190	5*
27. Head breadth	145	155	165	6	140	150	160	5*
28. Hand length	175	190	205	8	160	175	190	8*
29. Hand breadth	80	90	100	5	75	80	85	4*
30. Foot length	240	260	280	12	210	230	250	11
31. Foot breadth	85	95	105	5	75	85	95	5
32. Span	1640	1755	1870	70	1505	1610	1715	65*
33. Elbow span	795	860	925	38	720	785	850	41*
34. Vertical grip reach (standing)	2065	2205	2345	84	1875	2005	2135	79*
35. Vertical grip reach (sitting)	1210	1290	1370	50	1115	1185	1255	42*
36. Forward grip reach	730	795	860	38	680	735	790	34*

See notes on p. 83.

Table 4.35. Anthropometric estimates for Japanese adults (all dimensions in millimetres).

Dimension	Men				Women			
	5th %ile	50th %ile	95th %ile	SD	5th %ile	50th %ile	95th %ile	SD
1. Stature	1560	1655	1750	58	1450	1530	1610	48*
2. Eye height	1445	1540	1635	57	1350	1425	1500	47
3. Shoulder height	1250	1340	1430	54	1075	1145	1215	44
4. Elbow height	965	1035	1105	43	895	955	1015	36
5. Hip height	765	830	895	41	700	755	810	33
6. Knuckle height	675	740	805	40	650	705	760	33
7. Fingertip height	565	630	695	38	540	600	660	35
8. Sitting height	850	900	950	31	800	845	890	28
9. Sitting eye height	735	785	835	31	690	735	780	28
10. Sitting shoulder height	545	590	635	28	510	555	600	26
11. Sitting elbow height	220	260	300	23	215	250	285	20
12. Thigh thickness	110	135	160	14	105	130	155	14
13. Buttock–knee length	500	550	600	29	485	530	575	26
14. Buttock–popliteal length	410	470	510	31	405	450	495	26
15. Knee height	450	490	530	23	420	450	480	18
16. Popliteal height	360	400	440	24	325	360	395	21
17. Shoulder breadth (bideltoid)	405	440	475	22	365	395	425	18
18. Shoulder breadth (biacromial)	350	380	410	18	315	340	365	15
19. Hip breadth	280	305	330	14	270	305	340	20
20. Chest (bust) depth	180	205	230	16	175	205	235	18
21. Abdominal depth	185	220	255	22	170	205	240	20
22. Shoulder–elbow length	295	330	365	21	270	300	330	17
23. Elbow–fingertip length	405	440	475	20	370	400	430	17
24. Upper limb length	665	715	765	29	605	645	685	25
25. Shoulder–grip length	565	610	655	26	515	550	585	22
26. Head length	170	185	200	8	160	170	180	7
27. Head breadth	145	155	165	7	140	150	160	6
28. Hand length	165	180	195	10	150	165	180	9
29. Hand breadth	75	85	95	6	65	75	85	5
30. Foot length	230	245	260	10	210	225	240	9
31. Foot breadth	95	105	115	5	90	95	100	4
32. Span	1540	1655	1770	70	1395	1485	1575	56
33. Elbow span	790	870	950	48	715	780	845	41
34. Vertical grip reach (standing)	1805	1940	2075	83	1680	1795	1910	69
35. Vertical grip reach (sitting)	1105	1185	1265	49	1030	1095	1160	41
36. Forward grip reach	630	690	750	37	570	620	670	31

See notes on p. 83.

Table 4.36. Anthropometric estimates for Hong Kong Chinese adults (all dimensions in millimetres).

	Men				Women			
Dimension	5th %ile	50th %ile	95th %ile	SD	5th %ile	50th %ile	95th %ile	SD
1. Stature	1585	1680	1775	58	1455	1555	1655	60*
2. Eye height	1470	1555	1640	52	1330	1425	1520	57*
3. Shoulder height	1300	1380	1460	50	1180	1265	1350	51*
4. Elbow height	950	1015	1080	39	870	935	1000	41*
5. Hip height	790	855	920	41	715	785	855	42
6. Knuckle height	685	750	815	40	650	715	780	41
7. Fingertip height	575	640	705	38	540	610	680	44
8. Sitting height	845	900	955	34	780	840	900	37*
9. Sitting eye height	720	780	840	35	660	720	780	35*
10. Sitting shoulder height	555	605	655	31	510	560	610	29*
11. Sitting elbow height	190	240	290	31	165	230	295	38*
12. Thigh thickness	110	135	160	14	105	130	155	14
13. Buttock–knee length	505	550	595	26	470	520	570	30*
14. Buttock–popliteal length	405	450	495	26	385	435	485	29*
15. Knee height	450	495	540	26	410	455	500	27*
16. Popliteal height	365	405	445	25	325	375	425	29*
17. Shoulder breadth (bideltoid)	380	425	470	26	335	385	435	29*
18. Shoulder breadth (biacromial)	335	365	395	19	315	350	385	22
19. Hip breadth	300	335	370	22	295	330	365	21*(M)
20. Chest (bust) depth	155	195	235	25	160	215	270	34
21. Abdominal depth	150	210	270	36	150	215	280	39
22. Shoulder–elbow length	310	340	370	19	290	315	340	16*
23. Elbow–fingertip length	410	445	480	22	360	400	440	24*
24. Upper limb length	680	730	780	30	615	660	705	26
25. Shoulder–grip length	580	620	660	25	525	560	595	22
26. Head length	175	190	205	8	160	175	190	9
27. Head breadth	150	160	170	7	135	150	165	8
28. Hand length	165	180	195	9	150	165	180	9*
29. Hand breadth	70	80	90	5	60	70	80	5*
30. Foot length	235	250	265	10	205	225	245	11*
31. Foot breadth	85	95	105	5	80	85	90	4*
32. Span	1480	1635	1790	95	1350	1480	1610	80*
33. Elbow span	805	885	965	48	690	775	860	51
34. Vertical grip reach (standing)	1835	1970	2105	83	1685	1825	1965	86
35. Vertical grip reach (sitting)	1110	1205	1300	58	855	940	1025	51
36. Forward grip reach	640	705	770	38	580	635	690	32

See notes on p. 83.

Table 4.37. Anthropometric estimates for Indian adults (all dimensions in millimetres).

Dimension	Men				Women			
	5th %ile	50th %ile	95th %ile	SD	5th %ile	50th %ile	95th %ile	SD
1. Stature	1535	1640	1745	65	1415	1515	1615	61
2. Eye height	1440	1535	1630	59	1320	1420	1520	60
3. Shoulder height	1245	1345	1445	61	1140	1235	1330	57
4. Elbow height	950	1030	1110	48	875	950	1025	45
5. Hip height	780	855	930	46	710	780	850	43
6. Knuckle height	650	715	780	38	620	680	740	35
7. Fingertip height	560	620	680	36	525	590	655	38
8. Sitting height	795	850	905	33	730	790	850	35
9. Sitting eye height	685	740	795	32	635	690	745	33
10. Sitting shoulder height	590	640	690	29	465	515	565	31
11. Sitting elbow height	180	230	280	29	175	220	265	28
12. Thigh thickness	85	110	135	14	80	108	135	16
13. Buttock–knee length	510	560	610	29	490	540	590	29
14. Buttock–popliteal length	415	465	515	31	405	455	505	29
15. Knee height	465	515	565	29	425	470	515	26
16. Popliteal height	370	415	460	27	330	375	420	27
17. Shoulder breadth (bideltoid)	375	415	455	25	310	350	390	24
18. Shoulder breadth (biacromial)	325	355	385	18	285	315	345	17
19. Hip breadth	275	320	365	26	270	330	390	35
20. Chest (bust) depth	145	190	235	28	165	210	255	26
21. Abdominal depth	140	205	270	40	150	210	270	37
22. Shoulder–elbow length	310	340	370	19	280	310	340	17
23. Elbow–fingertip length	415	450	485	20	375	405	435	19
24. Upper limb length	680	735	790	33	610	665	720	32
25. Shoulder–grip length	575	625	675	30	515	565	615	29
26. Head length	175	185	195	7	160	170	180	7
27. Head breadth	135	145	155	6	125	135	145	5
28. Hand length	165	180	195	9	150	165	180	9
29. Hand breadth	70	80	90	5	65	70	75	4
30. Foot length	230	250	270	13	205	225	245	12
31. Foot breadth	80	90	100	6	75	85	95	5
32. Span	1565	1690	1815	77	1400	1515	1630	69
33. Elbow span	820	890	960	44	730	800	870	42
34. Vertical grip reach (standing)	1820	1940	2060	74	1680	1794	1910	69
35. Vertical grip reach (sitting)	1085	1175	1265	56	995	1083	1170	52
36. Forward grip reach	680	735	790	32	610	660	710	31

See notes on p. 84.

Table 4.38. Body weights (kg) of selelected adult populations.

	Men				Women			
Nationality	5th %ile	50th %ile	95th %ile	SD	5th %ile	50th %ile	95th %ile	SD
British (19–65 years)[a]	55·3	74·5	93·7	11·7	44·1	62·5	80·9	11·2
USA (19–65 years)[a]	55·2	78·4	101·6	14·1	40·5	64·7	88·9	14·7
French	58·2	73·2	94·6	11·0	46·5	58·0	78·0	10·0
German	59·4	76·2	95·7	11·7	—	—	—	—
Swedish[a]	—	—	—	—	48·3	59·3	70·2	6·7
Japanese[a]	46·1	60·2	74·3	8·6	39·8	51·3	62·8	7·0
Hong Kong Chinese	46·6	59·9	75·3	8·6	38·5	47·1	61·8	7·2
Indian	—	49·2	—	—	—	43·5	—	—

[a]Calculated from parameters of normal distribution.

See notes on p. 84.

Table 4.39. Body weights (kg) of British 3–18-year-olds.

	Boys				Girls			
Age (years)	5th %ile	50th %ile	95th %ile	SD	5th %ile	50th %ile	95th %ile	SD
3	13·5	16·5	19·5	2·0	12·5	15·5	19·0	1·9
4	14·5	18·0	21·5	2·1	13·5	17·5	21·5	2·4
5	15·5	20·0	24·5	2·8	15·0	19·5	24·0	2·8
6	16·0	22·0	28·0	3·6	16·5	21·5	27·0	3·2
7	18·5	24·5	30·0	3·4	17·5	24·0	30·5	3·9
8	20·0	27·0	34·0	4·3	19·5	27·0	35·0	4·7
9	22·0	30·0	37·5	4·6	20·5	30·0	39·5	5·8
10	23·5	33·5	43·0	5·9	22·5	43·0	45·5	6·9
11	25·5	36·5	48·0	6·9	24·5	38·0	51·5	8·3
12	27·5	40·5	54·0	7·9	28·5	42·0	56·0	8·3
13	30·5	45·5	61·0	9·3	32·5	47·5	62·5	9·2
14	35·0	52·0	69·0	10·3	38·0	52·0	65·5	8·5
15	40·5	57·0	73·5	10·0	40·5	54·5	68·6	8·6
16	46·5	61·5	77·0	9·3	43·0	55·5	67·5	7·5
17	51·0	65·5	80·5	8·9	43·5	56·0	69·0	7·8
18	53·5	67·0	80·5	8·2	45·0	56·5	68·0	7·0

Data quoted from Department of Education and Science (1985).

See notes on p. 84.

Chapter 5
Anthropometry of special regions of the body

The purpose of this chapter is to present static anthropometric data concerning four regions of the body which are important in certain areas of design. The tables have been compiled to represent the general population of Britain, aged 19–65 years, i.e., equivalent to Table 4.1 (relevant items from which have been repeated for the users' convenience). Where estimation has been necessary, scaling has been based on the nearest equivalent known dimension, e.g., head length, foot breadth, etc.

5.1. *The head and face*

The design of protective headgear, facemasks, etc., has prompted military researchers to collect extensive anthropometric data concerning the head and face. Table 5.1 shows that some military groups are very similar to civilian estimates; Table 5.2 has been prepared by scaling the military sources down or up where necessary. The data will be appropriate for the location of eye-, ear- or

Table 5.1. Head data (mm)—comparison of sources.

	Men				Women			
	Head length		Head breadth		Head length		Head breadth	
Source	Mean	SD	Mean	SD	Mean	SD	Mean	SD
1. British civilian estimates (see Table 4.1)	195	8	153	6	178	7	143	6
2. British infantrymen (Godderson and Beebee 1976)	195	7	154	6				
3. US Airforce (NASA 1978)	196	7	153	6	184	7	145	6

[a]These dimensions have been rounded to the nearest 1 mm (not 5 mm).

Table 5.2. Anthropometric estimates for the head and face (all dimensions in millimetres).

Dimension	Men				Women			
	5th %ile	50th %ile	95th %ile	SD	5th %ile	50th %ile	95th %ile	SD
1. Head length	180	195	205	8	165	180	190	7
2. Head breadth	145	155	165	6	135	145	155	6
3. Maximum diameter from chin	240	255	265	8	225	235	245	7
4. Chin to top of head[a]	205	225	240	11	200	220	240	11
5. Ear to top of head[b]	115	125	135	6	110	125	135	8
6. Ear to back of head[b]	90	100	115	7	85	100	115	9
7. Bitragion breadth[a]	125	135	145	6	120	130	135	5
8. Eye to top of head[c]	105	115	125	7	100	115	130	9
9. Eye to back of head[c]	160	170	185	8	145	160	175	10
10. Interpupillary breadth	55	60	70	4	55	60	65	4
11. Nose to top of head[d]	130	150	165	10	125	145	165	12
12. Nose to back of head[d]	205	220	235	9	190	205	220	10
13. Mouth to top of head[e]	165	180	195	9	155	170	190	11
14. Lip length[f]	40	50	55	5	35	45	50	4

[a] Measured to menton, i.e., the palpable bony tip of the chin.
[b] Measured to tragion, i.e., the flap of cartilage which covers the actual orifice of the ear.
[c] Measured to ectocanthus, i.e., the outer angle of the eye.
[d] Measured to pronasale, i.e., the tip of the nose.
[e] Measured to stomion, i.e., the midpoint of the closed lips.
[f] Measured with the lips relaxed.

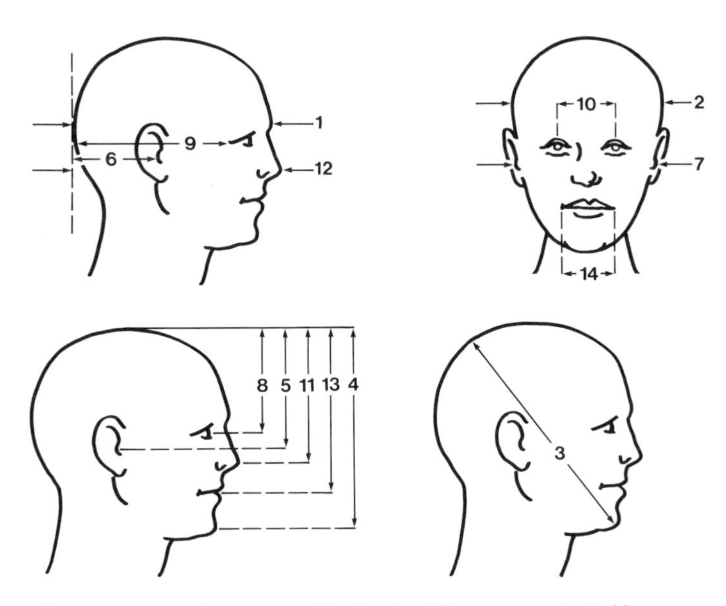

Figure 5.1. Anthropometry of the head and face, as given in Table 5.2.

mouthpieces. Such problems are increasingly common in 'new technology' products. The designer of protective equipment (facemasks, etc.) should remember that these data are only estimated—in situations where the health and safety of the user is at issue extensive user trials are essential.

All measurements (except 2, 3, 10 and 14) are made perpendicular either to the horizontal plane which touches the top of the head or to the vertical plane which touches the back of the head (see Figure 5.1).

5.2. *The back*

The human vertebral column (backbone) consists of 24 movable bony vertebrae separated by deformable hydraulic pads of fibrocartilage known as intervertebral discs. (Up to 10% of people possess a greater or lesser number of vertebrae but these 'anomalies' seem to have little functional consequence.) The column is surmounted by the skull, and rests upon the sacrum which is firmly bound to the hip bones at the sacro-iliac joints. The vertebrae can be naturally grouped into seven cervical (in the neck), twelve thoracic (to which the ribs are attached) and five lumbar (in the small of the back, between the ribs and the pelvis). The spine is a flexible structure, the configuration of which is controlled by many muscles and ligaments (Figure 5.2).

In the upright standing or sitting position the well-formed human spine presents a sinuous curve when viewed in profile. The painter Hogarth (1753) called this curve 'the line of beauty'—very appropriately in my opinion. The cervical region is concave (to the rear), the thoracic region convex and the lumbar region again concave. A concavity is sometimes known as a 'lordosis' and a convexity as a

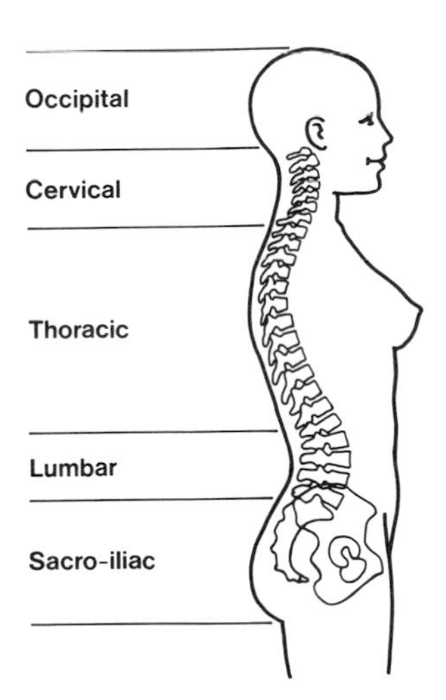

Figure 5.2. The well-formed human spine presents a sinuous curve when viewed in profile.

Table 5.3. Anthropometric estimates for the back (all dimensions in millimetres).

Dimension	Men 5th %ile	Men 50th %ile	Men 95th %ile	Men SD	Women 5th %ile	Women 50th %ile	Women 95th %ile	Women SD
1. Sitting height[a]	850	910	965	36	795	850	910	35
2. Occipital height (sitting)[b]	765	830	900	38	710	770	825	35
3. Nape height (sitting)[c]	660	725	785	38	605	660	720	35
4. C7 height (sitting)[d]	605	660	710	33	565	615	665	31
5. Scapular height (sitting)[e]	405	445	480	22	380	415	450	21
6. Lumbar height (sitting)[f]	195	240	285	26	195	230	265	22
7. Sacral height (sitting)[g]	125	165	200	23	130	165	200	21
8. Shoulder breadth (bideltoid)[a]	420	465	510	28	355	395	435	24
9. Chest breadth[h]	275	310	345	21	235	265	295	18
10. Elbow–elbow breadth	370	450	530	49	320	385	455	41
11. Waist breadth	250	290	330	24	200	230	260	18
12. Hip breadth[a]	310	360	405	29	310	370	435	38

[a]Quoted from Table 4.1.
[b]From seat to the most posterior point of the back of the head.
[c]From seat to the midpoint of the concavity of the neck.
[d]From seat to the palpable spine of the seventh cervical vertebra (vertebra prominens).
[e]From seat to the lower tip of the shoulder blade—for practical purposes equivalent to the midpoint of the thoracic convexity.
[f]From seat to the midpoint of the lumbar concavity.
[g]From seat to the posterior superior iliac spine—a bony landmark appropriately known as the 'dimple of Venus'.
[h]Measured at the level of the nipple.

Figure 5.3. Anthropometry of the back, as given in Table 5.3.

'kyphosis'. These are enclosed by the convexities of the occiput (back of the head) above with the sacro-iliac region and buttocks below; making five curves in all. The design of seats (particularly those in transport systems) requires data concerning these curves. It is surprising how little of such data exists. Table 5.3 is patched together using Branton (1984), Martin (1960) and NASA (1978) as shape sources for dimensions 2–7 with the sitting height estimates of Table 4.27 as a size reference. Dimensions 9–11 were scaled from stature using shape references from NASA (1978).

The curves of the human spine are in reality highly variable; Branton (1984) has described this variability.

5.3. *The hand*

The design of handgrips and hand-operated controls presents important ergonomic problems. Consequently, the anthropometry of the hand has been well investigated, important sources including Kember *et al.* (1981) and Davies *et al.* (1980) for British industrial workers, Gooderson *et al.* (1982) for British military personnel and Garret (1971) for US service personnel. Table 5.4 compares these sources with the estimates of Table 4.2. In Table 5.5 dimensions 1 and 12 are from Table 4.27, dimensions 2–11, 13 and 15 are quoted from Kember *et al.* (1981), dimensions 16, 17 and 19 are quoted from Gooderson *et al.* (1982), dimension 20 is quoted directly from Davies *et al.* (1980) for women and estimated by scaling for men and dimension 18 is scaled down from Garret (1971). Note that in these tables data are presented to the nearest 1 mm.

Table 5.4. Hand data—comparison of sources.

	Men				Women			
	Hand length		Hand breadth		Hand length		Hand breadth	
Source	Mean	SD	Mean	SD	Mean	SD	Mean	SD
1. British civilian estimates (from Table 4.1)	189	10	87	5	174	9	76	4
2. British industrial workers (Kember *et al.* 1981)	191	8	88	4	174	7	77	4
3. British industrial workers (Davies *et al.* 1980 a)	—	—	—	—	174	9	77	5
4. British military personnel (Gooderson *et al.* 1982)	191	9	86	4	176	8	78	4
5. US service personnel (Garret 1971)	197	9	90	4	179	8	77	4

Table 5.5. Anthropometric estimates for the hand (all dimensions in millimetres).

Dimension	Men 5th %ile	Men 50th %ile	Men 95th %ile	Men SD	Women 5th %ile	Women 50th %ile	Women 95th %ile	Women SD
1. Hand length	173	189	205	10	159	174	189	9
2. Palm length	98	107	116	6	89	97	105	5
3. Thumb length	44	51	58	4	40	47	53	4
4. Index finger length	64	72	79	5	60	67	74	4
5. Middle finger length	76	83	90	5	69	77	84	5
6. Ring finger length	65	72	80	4	59	66	73	4
7. Little finger length	48	55	63	4	43	50	57	4
8. Thumb breadth (IPJ)[a]	20	23	26	2	17	19	21	2
9. Thumb thickness (IPJ)	19	22	24	2	15	18	20	2
10. Index finger breadth (PIPJ)[b]	19	21	23	1	16	18	20	1
11. Index finger thickness (PIPJ)	17	19	21	1	14	16	18	1
12. Hand breadth (metacarpal)	78	87	95	5	69	76	83	4
13. Hand breadth (across thumb)	97	105	114	5	84	92	99	5
14. Hand breadth (minimum)[c]	71	81	91	6	63	71	79	5
15. Hand thickness (metacarpal)	27	33	38	3	24	28	33	3
16. Hand thickness (including thumb)	44	51	58	4	40	45	50	3
17. Maximum grip diameter[d]	45	52	59	4	43	48	53	3
18. Maximum spread	178	206	234	17	165	190	215	15
19. Maximum functional spread[e]	122	142	162	12	109	127	145	11
20. Minimum square access[f]	56	66	76	6	50	58	67	5

[a] IPJ is the interphalangeal joint, i.e., the articulations between the two segments of the thumb.
[b] PIPJ is the proximal interphalangeal joint, i.e., the finger articulation nearest to the hand.
[c] As for dimension 12, except that the palm is contracted to make it as narrow as possible.
[d] Measured by sliding the hand down a graduated cone until the thumb and middle fingers only just touch.
[e] Measured by gripping a flat wooden wedge with the tip end segments of the thumb and ring fingers.
[f] The side of the smallest equal aperture through which the hand will pass.

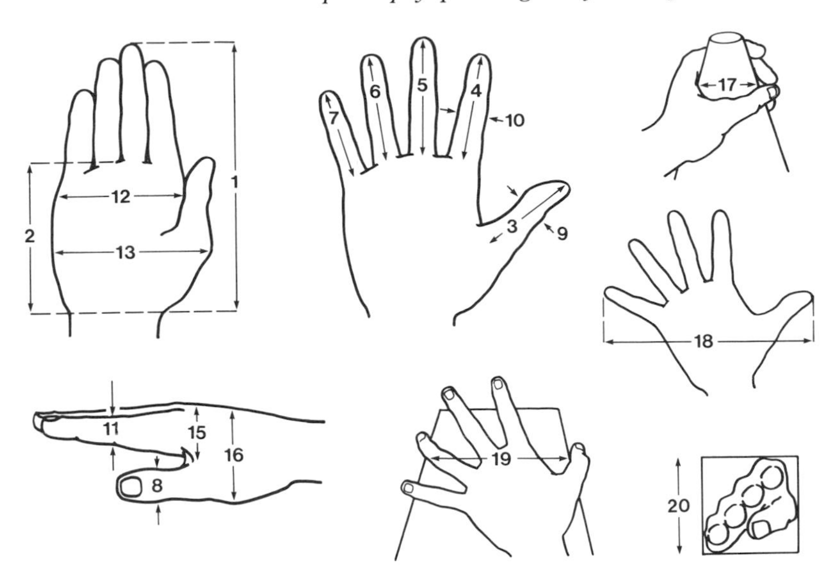

Figure 5.4. Anthropometry of the hand, as given in Table 5.5.

5.4. *The foot*

Some additional information concerning the foot may be of use in the design of pedals, etc. Table 5.6 gives estimates for adult foot dimensions using NASA (1978) and Hansen and Cornog (1958) as reference sources and scaling to the foot length and breadth of Table 4.1. These are dimensions of the unshod foot.

Table 5.6. Anthropometric estimates for the foot (all dimensions in millimetres).

Dimension	Men				Women			
	5th %ile	50th %ile	95th %ile	SD	5th %ile	50th %ile	95th %ile	SD
1. Foot length	240	265	285	14	215	235	255	12
2. Instep length[a]	175	190	210	11	160	175	190	10
3. Fifth toe length	195	215	235	12	180	195	210	10
4. Foot breadth	85	95	110	6	80	90	100	6
5. Heel breadth	60	70	75	5	50	55	65	6
6. Ankle height[b]	60	70	85	7	55	65	75	6
7. Dorsal arch height[c]	70	80	90	6	60	70	80	5
8. Heel–ankle horizontal[b]	50	55	60	5	45	50	55	5

[a]Measured to the medial prominence of the ball of the foot.
[b]Measured to the lateral malleolus of the fibula, i.e., the approximate centre of rotation of the ankle joints.
[c]Measured to the point where the top of the foot meets the front of the leg.

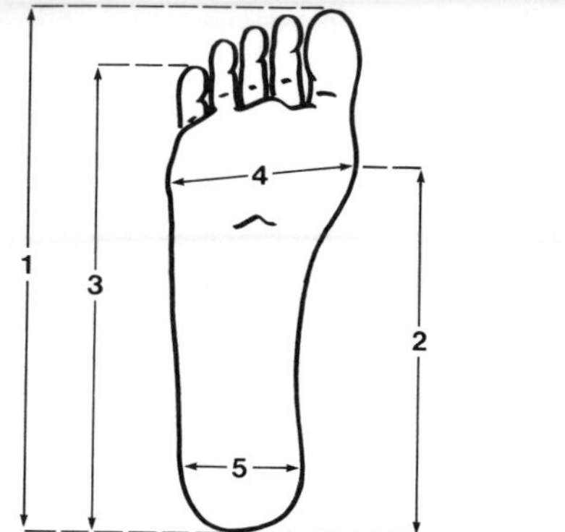

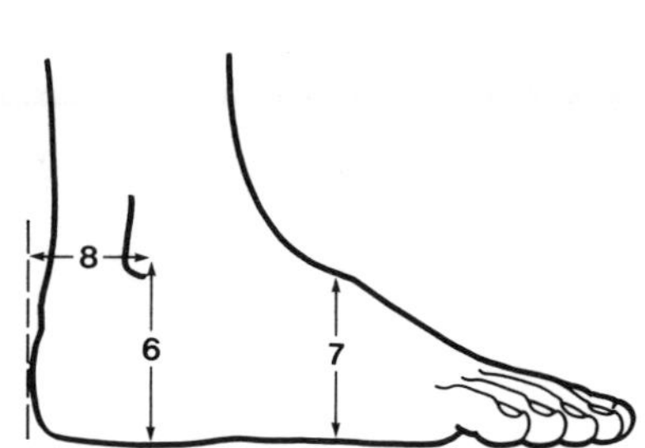

Figure 5.5. Anthropometry of the foot.

Chapter 6
Body segment parameters—Newtonian anthropometry

Suppose we wish to analyse the biomechanical stresses to the body in a particular posture, or we wish to make a mathematical model of a dynamic activity such as locomotion, an industrial handling task or a sporting skill—a different type of datum from those we have encountered so far will be required. Since the purpose of such data is to allow the numerical application of Newton's Laws of Motion to the human body or its parts, the term 'Newtonian anthropometry' has been coined for its description (Grieve and Pheasant 1982).

In theory, perhaps, the solution of biomechanical problems would require a knowledge of the motion of each individual particle within the body. The net forces existing between these particles, together with those exerted upon the outside world, could then be determined. This is not a practical proposition: the collection of such data presenting insurmountable obstacles both to instrumentation and to human comprehension. As a compromise we may divide the body into as small a number of segments as possible; compatible with the requirement that an analysis of their motions should give an acceptably accurate solution to the problems under consideration. (The analytical techniques of biomechanics are beyond the scope of the present volume. Considerable literature exists on this subject—the reader is referred to Plagenhof (1971), Miller and Nelson (1973), Winter (1979), Grieve and Pheasant (1982) and Chaffin and Andersson (1984).)

6.1. *Linked-body Man*

The segments of the body are often referred to, by analogy with machines, as 'linkages'. Hence, the thigh linkage extends from the hip joint to the knee joint, the forearm from the elbow joint to the wrist joint, etc. The terms 'segment' and 'linkage' are more or less interchangeable and the underlying conceptual model is often referred to as the 'Linked-body Man'. In order to proceed we must assume the following:

(*i*) The body may be divided into a (small) number of finite rigid linkages, the lengths and inertial properties of which are constant for any particular individual.

(*ii*) The linkages articulate at frictionless joints, the centres of rotation of which are constant.

These statements may be challenged both in detail and in essence. The trunk, for example, if taken as a whole, is far from rigid. We may, of course, subdivide the trunk into several segments but it then becomes difficult to attribute the weight of the viscera to a certain location. The centre of rotation of the shoulder (glenohumeral) joint is far from constant—as the shoulder girdle moves on the thorax. Again, we can, if necessary, deal with this problem by introducing additional linkages.

The complexity of our calculations will increase with the number of linkages involved, but experience in diverse areas of application suggests that the simplest of biomechanical models are capable of yielding useful insights into practical problems. For present purposes each limb will be divided into three segments. Data will also be given for the head, neck and trunk as a unit, together with its subdivisions. Because of the importance of calculations involving the lumbosacral joint, the trunk will again be divided into a part which lies above this level and a part below.

Historically, the parameters of body segments have been determined by the careful and systematic dismembering of cadavers. The collection of data by this means commenced with the work of German anatomists such as Harless, Braune and Fischer at the end of the last century. The definitive study of the subject undoubtedly remains that of Dempster (1955), whose data have been quoted and requoted and applied in areas ranging from astronautics to industrial safety, automobile crash dummies and sport. It is sobering to record that Dempster's work was based upon only eight specimens. Subsequent studies have increased this number to 65 in all (Reynolds 1978) but differences in technique make comparisons difficult. It is possible to supplement cadaveric data with measurements made on the living (Drillis and Contini 1966, Miller and Nelson 1973, Roebuck *et al.* 1975). Information, however, remains very sparse and, considering the increasing importance of the computer models into which these data are incorporated, a definitive study is long overdue.

The present dataset derives originally from Dempster (1955) but it has subsequently passed through several hands, the most recent of which are Reynolds (1978) and Winter (1979). Its accuracy, quite simply, is unknown and impossible to estimate.

6.2. *Linkage lengths*

Table 6.1 gives the average length of the important linkages of the body (as shown in Figure 6.1) expressed as a percentage of stature. Major limb segments are based on the re-analysis of Dempster by Reynolds (1978). These have been considerably modified using the bodily proportional data upon which the tables of Chapter 4 are

Table 6.1. Average link lengths as a percentage of stature.

Segment	Men	Women
1. Upper arm	17·4	17·2
2. Forearm	15·6	14·9
3. Hand	10·9	10·8
4. Thigh	24·3	24·2
5. Shank	23·6	23·0
6. Foot	15·2	14·7
7. Ankle above sole	4·2	4·1
8. Ankle in front of heel	3·3	3·2
9. Hip–lumbosacral joint	5·7	5·7
10. Hip–shoulder joint	28·8	30·4
11. Hip–C7	33·4	34·0
12. Hip–vertex	47·9	48·7
13. C7–vertex	14·5	14·7
14. Hip above the SRP	4·3	4·3
15. Hip in front of the SRP	7·0	8·2
16. Transverse shoulders	21·9	21·2
17. Transverse hips	9·9	10·9

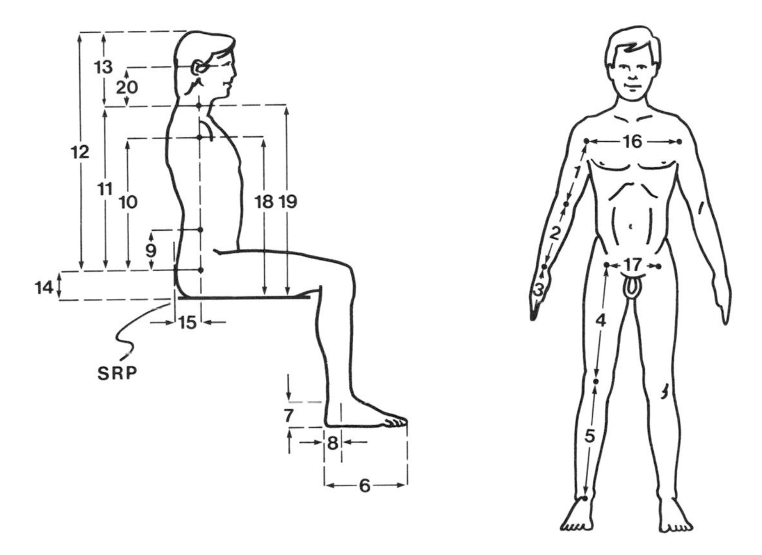

Figure 6.1. Anthropometry of body linkages, as given in Tables 6.1 and 6.2.

based in order to give a model which is complete, internally consistent and numerically compatible with the rest of this book. It should be noted that we cannot obtain a 95th %ile thigh length, for example, by taking the appropriate percentage of the 95th %ile of stature. That calculation would yield the thigh length of a person of 95th %ile stature and average bodily proportions—which is not quite the same thing.

Table 6.2. Estimated linkage dimensions (mm) for British adults.

Segment	Men				Women			
	5th %ile	50th %ile	95th %ile	SD	5th %ile	50th %ile	95th %ile	SD
1. Upper arm	275	305	335	17	250	275	300	15
2. Forearm	245	270	295	15	220	240	260	13
3. Hand	175	190	205	10	160	175	190	9
4. Thigh	385	425	465	23	355	390	425	22
5. Shank	375	410	445	22	340	370	405	20
6. Foot	240	265	285	14	215	235	255	12
7. Ankle above sole	70	75	80	4	60	65	70	4
8. Ankle in front of heel	50	55	60	3	45	50	55	3
9. Hip–lumbosacral joint	90	100	110	6	80	90	100	5
10. Hip–shoulder joint	455	500	540	26	445	490	535	26
11. Hip–C7	530	580	630	30	495	545	595	29
12. Hip–vertex	765	835	905	42	720	785	850	38
13. C7–vertex	230	255	280	14	220	240	260	13
14. Hip above the SRP	70	75	80	4	65	70	75	4
15. Hip in front of the SRP	110	120	130	7	120	130	140	7
16. Transverse shoulders	345	380	410	20	310	340	370	19
17. Transverse hips	155	170	185	9	160	175	190	10
18. Shoulder–SRP	525	575	625	30	510	560	610	29
19. C7–SRP	600	655	710	34	560	615	670	32
20. C7–eye	120	135	150	8	115	125	135	7

In Table 6.2, therefore, the average values have been calculated directly from the percentages, the standard deviations have been calculated using Equations 2.47 and 2.48 and the percentiles have been calculated from the distributions which these parameters define. The figures quoted are for the standard reference population of British adults aged 19–65 years (i.e., equivalent to Table 4.1).

6.3. Inertial characteristics

The inertial characteristics of the segments are presented in Tables 6.3 and 6.4. The segmental weights of Table 6.2 are quoted from Reynolds (1978) as are some of the locations of segmental centres of gravity. The remainder are the present author's calculations based on the original data of Dempster (1955) together with the dimensions of Table 4.1. The centre of gravity of a segment may be defined as that point through which its weight may be assumed to act. Its location is important in the calculation of postural stress (see Section 8.2).

The moment of inertia of a body (I) is a measure of its resistance to rotation (as its mass is for translation—Newton's Second Law). Hence,

$$\tau = I\omega \tag{6.1}$$

where τ is the torque and ω the angular acceleration.

Table 6.3. Weights and centres of gravity of body segments.

Segment	Weight (% body weight)	Location of centre of gravity
1. Head and neck	8·4	57% of distance from C7 to vertex
1a. Head	6·2	20 mm above tragion
2. Head and neck and trunk	58·4	40% of distance from hip to vertex
2a. Trunk	50·0	46% of distance from hip to C7
2b. Trunk above lumbosacral joint	36·6	63% of distance from hip to C7
2c. Trunk below lumbosacral joint	13·4	Approximately at the hip joint
3. Upper arm	2·8	48% of distance from shoulder to elbow joints
4. Forearm	1·7	41% of distance from elbow to wrist joints
5. Hand	0·6	40% of hand length from wrist joint (at the centre of an object gripped)
6. Thigh	10·0	41% of distance from hip to knee joints
7. Lower leg	4·3	44% of distance from knee to ankle joints
8. Foot	1·4	47% foot length forward from the heel (half height of ankle joint above the ground)–midway between ankle and ball of foot at the head of metatarsal III

Table 6.4. Radius of gyration of body segments.[a]

	About centre of gravity	About proximal joint	About distal joint
1. Head, neck and trunk	30·2	65·7	—
2. Head and neck	25·6	62·4	—
3. Upper arm	32·2	57·8	61·2
4. Forearm	30·3	51·0	66·3
5. Hand	21·6	44·6	—
6. Thigh	32·3	52·2	67·2
7. Lower leg	30·2	53·4	63·6
8. Foot	27·8	54·6	—

[a]Data presented as percentages of segment length—for head, neck and trunk this is hip to vertex, for head and neck this is C7 to vertex.

The moment of inertia of a body must always be specified with respect to a particular axis of rotation. For reasons which a moment's thought or reference to a physics book should make clear, the lowest moment of inertia (I_0) of a body in a particular plane will be about an axis perpendicular to that plane passing through the centre of gravity. The moment of inertia about any other parallel axis (I_x) may be calculated by the equation:

$$I_x = I_0 + md^2 \tag{6.2}$$

where m is the mass of the body and d is the perpendicular distance between the axes. This relation is called the 'parallel axis theorem'.

For convenience the moment of inertia of a body segment is usually converted to a radius of gyration (k) defined by the equation:

$$k = \sqrt{I/m} \qquad (6.3)$$

This may now be expressed as a percentage of linkage length. Again, the smallest radius of gyration will be the one about the centre of gravity.

The data of Table 6.4 are all for rotation in a sagittal plane, i.e., for movements of flexion and extension. The radii of gyration about the centres of gravity are taken from Winter (1979) who, in turn, derived them via various intermediaries from Dempster (1955)—minor modifications have been made for different definitions of linkage length. Bodily segments rarely rotate about their centres of gravity—more usually they rotate about the proximal articulation, less usually about the distal. The radii of gyration about these axes have been calculated using the parallel axis theorem and the data of Table 6.3.

Chapter 7
The dynamic anthropometry of clearance and reach

In this chapter we shall give further consideration to reach and clearance. Certain of the data of Chapter 4 are concerned with these matters but we shall now consider not only static dimensions but dynamic anthropometrics as well (see Section 2.3). Dynamic data are generally obtained in special purpose studies which often include factors which render the data inapplicable outside the original area of concern—hence, reach data obtained for the design of aircraft cockpits may be inappropriate to motor-cars or industrial workstations (due to differences in seating, restraint belts, etc.). It is often possible to circumvent these difficulties by mathematical modelling (generally using the linkage data of Table 6.2)—we shall call this the pseudo-dynamic approach to anthropometry.

Unless otherwise specified all the data in this chapter have been scaled to correspond to the standard reference population of British adults aged 19–65 years (i.e., equivalent to Table 4.1).

7.1. *Clearance*

Table 7.1 and Figure 7.1 present clearance data for a variety of working positions, derived from a variety of sources (Damon *et al.* 1966, Van Cott and Kinkade 1972, Department of Defense 1981) and scaled, as far as possible, to match the standard population. The maximum breadth and depth of the body are overall measurements taken at the widest or deepest point wherever this occurs. The male data, based on US servicemen, exceed any relevant dimensions in Table 4.1 and have been quoted direct.

Fruin (1971), in the context of an account of pedestrian movement and flow, introduced the concept of the body ellipse. In plan view the space occupied by the human body may be approximately described by an ellipse—the long and short axes of which are determined by its maximum breadth and depth. Taking the 95th %ile male data from Table 7.1, and allowing a generous 25 mm all round for clothes, the long and short axes of our ellipse become 630 and 380 mm, respectively. Figure 7.2 shows this ellipse. To give us some idea of 'elbow room' a circle has been drawn around the ellipse. The diameter of this circle is the 95th %ile male elbow span (1020 mm). Two further circles, the diameters of which are deter-

Table 7.1. Clearance dimensions in various positions (all dimensions in millimetres).

Dimension	Men 5th %ile	Men 50th %ile	Men 95th %ile	Men SD	Women 5th %ile	Women 50th %ile	Women 95th %ile	Women SD
1. Maximum body breadth	480	530	580	30	355	420	485	40
2. Maximum body depth	255	290	325	22	225	275	325	30
3. Kneeling height	1210	1295	1380	51	1130	1205	1285	45
4. Kneeling leg length	620	685	750	40	575	630	685	32
5. Crawling height	655	715	775	37	605	660	715	33
6. Crawling length	1215	1340	1465	75	1130	1240	1350	66
7. Buttock–heel length	985	1070	1160	53	875	965	1055	55

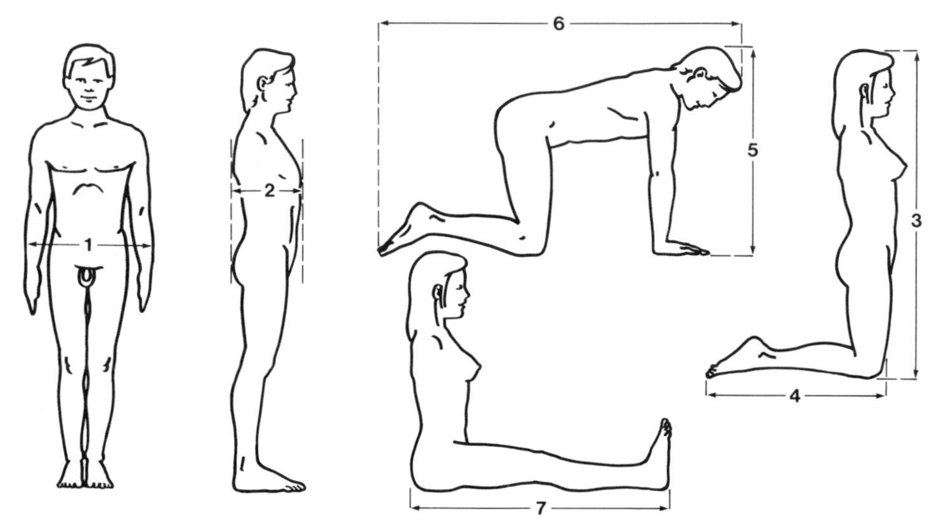

Figure 7.1. Clearance dimensions in various positions as given in Table 7.1.

mined by the arm span of a 5th %ile woman and a 95th %ile man, complete a first simple analysis of space requirements.

It is sometimes necessary to consider the minimum clearance required to give personal access or passageway in severely constrained circumstances. Table 7.2 gives the minimum dimensions for hatches or similar apertures which will just allow one person to pass through. These data are quoted from MIL-STD-1472C (Department of Defense 1981) and, since they are deemed to be adequate for US service personnel (a bulky population), they will presumably be suitable for most other target groups. (Although it is worth noting that other sources quote different figures—Damon *et al.* (1966), Van Cott and Kinkade (1972), and elsewhere.) In the case of emergency exits and escape hatches we should expect speed of passing through to be a function of aperture size up to some critical dimension at which no

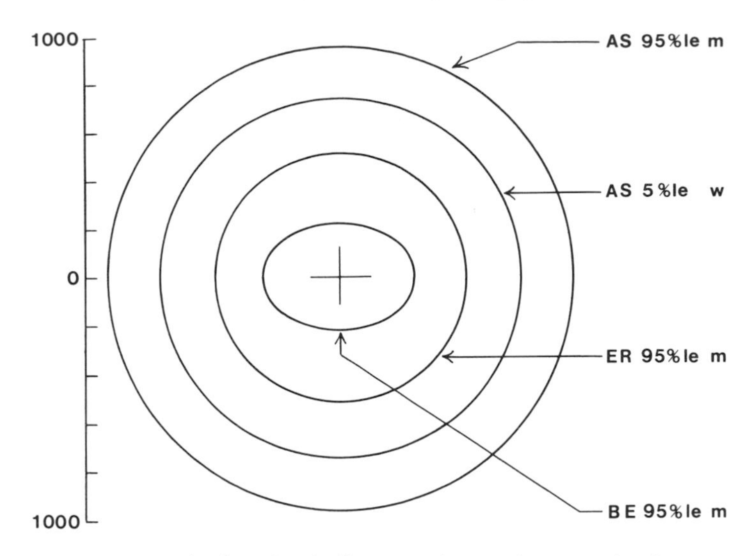

Figure 7.2. *Body ellipse (BE), elbow room (ER) and arm span (AS). See text for details.*

further improvement was possible. Roebuck and Levendahl (1961) studied the emergency exits of aircraft and found that speed levelled off at a door width of around 510 mm (unless steps were also involved, in which case greater widths were optimal).

Minimum dimensions for passageways in situations of limited access, such as tunnels and catwalks, etc., are given in Table 7.3 (quoted from Damon *et al.* (1966) with slight modifications, introduced for conformity with figures given elsewhere).

Wilson *et al* (1980) published an extensive review of space requirements in the home; which was subsequently summarized in a design guide by Noble (1982). The data of Table 7.4 are quoted directly. Table 7.5, however, is a summary of numerous separate recommendations which are made for circulation requirements around domestic obstacles such as chairs, tables and worktops.

Table 7.2. *Minimum acceptable dimensions for hatches giving whole body access.*

For sideways access through a rectangular aperture in a vertical surface (i.e., wall)	
Light clothing	660 mm high × 760 mm wide
Bulky clothing	740 mm high × 860 mm wide
For top or bottom access through a rectangular aperture in a horizontal surface (i.e., floor/cceiling)	
Light clothing	330 mm deep × 580 mm wide
Bulky clothing	410 mm deep × 690 mm wide
For a circular aperture	
Minimum diameter	760 mm

After Department of Defense (1981).

Table 7.3. Minimum dimensions for passageways in areas of restricted access.

	Height (mm)	Width (mm)
Walking[a]		
Upright	1955[b]	
Stooped	1600	
Straight ahead		630
Crabwise (sideways)		380
Crawling on hands and knees	815	630
Crawling prone[c]	430	630

[a]For walking a trapezoidal space which is 630 mm wide at shoulder height and 145 mm wide at floor level will suffice.
[b]Stature of a 99th %ile man wearing shoes and protective helmet.
[c]For prone crawling a width of 1015 mm is preferable to allow for lateral elbow movements.
Data based on Damon *et al.* (1966), with modifications.

Table 7.4. Circulation requirements in halls and corridors.

	Width (mm)
One person walking	650
One person carrying a tray or similar load	900
One person with a walking frame	1000
Two persons walking abreast	1350
One person walking, another pressed against a wall	1000
Two persons passing crabwise	900

Data from Noble (1982).

Table 7.5. Clearance required for walking between obstacles.

	Straight ahead (mm)	Crabwise (mm)
Between two walls or other obstacles greater than 1000 mm in height	650	400
Between one wall (or obstacle higher than 1000 mm) and one lower than 1000 mm	600	400
Between two obstacles lower than 1000 mm	550	350

7.2. *Reach—the workspace envelope*

Consider what happens when you reach your arm forwards. Firstly, you raise your upper limb through 90°—this is achieved principally by a rotational movement of the arm in its socket on the shoulder blade or scapula (i.e., flexion of the gleno-humeral or true shoulder joint; see Figure 6.2). It is, however, impossible to make such a movement without a small shift in the scapula's position on the chest wall. (Anatomists call this interaction 'scapulo-humeral rhythm'). You have now reached the position in which the 'static' dimensions of forward reach, dimension 36 in the tables of Chapter 4, would be measured—if your shoulder blades had been touching a wall at the outset they would still be doing so. As you reach farther forwards from this basic starting point, several new movements occur. Your whole shoulder girdle is thrust forwards (protracted) and you begin to incline your trunk forwards by flexion of the hip joint and spine. What determines the final limit of your forward reach? Try it and you will quickly discover that it is the tendency to topple over as the horizontal position of your centre of gravity reaches the limit of the base of support of your feet. This in turn can be modified by pushing the pelvis backwards as a counterbalance, etc. Suitable increments to the basic dimension of forward grip reach are given in Table 7.6.

Table 7.6. Increments to forward grip reach (all dimensions in millimetres).

Dimension	Men 5th %ile	Men 50th %ile	Men 95th %ile	Women 5th %ile	Women 50th %ile	Women 95th %ile
Basic dimension						
Forward grip reach[a]	720	780	835	650	705	755
Increments						
For a pinch grip (to the thumbtip)	35	40	40	30	35	40
For fingertip operation	105	115	125	95	105	115
For a forward thrust of the shoulder[b]	115	130	150	95	115	140
For 10° of trunk inclination[c]	80	85	95	75	85	95
For 20° of trunk inclination[c]	155	170	185	150	170	185
For 30° of trunk inclination[c]	230	250	270	225	245	270

[a]Quoted directly from Table 4.1.
[b]Calculated from data in Department of Defense (1981).
[c]Calculated from the linkage data in Table 6.2.

Dynamic reach may best be characterized by the three-dimensional co-ordinates of a volume of space. Such a volume is referred to as a 'workspace envelope' (or more grandly as a 'kinetosphere'). Since standing reach is essentially a matter of body equilibrium, the envelope will be modified by any factor which affects this. A weight in the hands will diminish reach. Grieve and Pheasant (1982) reported experiments showing how reach was increased by increasing the footbase and diminished by placing an obstacle behind the subject to limit the activities of counterbalancing.

Several studies of the workspace envelope of the sitting person have been published. That of Kennedy (1964) has been particularly widely quoted (Damon *et al.* 1966, Van Cott and Kinkade 1972, NASA 1978). The reader should note that all reach envelopes are highly specific to the situation in which they were measured. The data of Kennedy (1964) were measured in an aircraft seat with the subjects securely strapped in—had the seat or the restraints been otherwise the results would have been numerically different.

ISO 3958 copes with this problem by tabulating separate reach envelopes for different types of automobile; which are specified in terms of a 'package factor'. (This is a figure which is calculated from a formula involving a number of dimensions of the driver's workspace—see Section 13.1). Table 7.7 is derived from ISO 3958 with considerable modifications. Data for the package factor of a heavy goods vehicle were used since the relatively upright seat configuration will be most relevant for general purposes. The original survey was conducted wearing non-extensible seat belts and separate figures were quoted for 'outboard' and 'inboard' reaches—only the latter were used since they would be less affected by the belts. The original source uses the 'H point' (see Section 13.1) as the origin of the reach envelope and measures to a pinch grip. In the interests of uniformity with the remainder of this book, increments were added to shift the origin to the seat reference point (SRP) and a deduction was made for a full grip. The original data were obtained on a sample of subjects representative of the driving population of the USA; it is probable that the tabulated reaches are very slightly greater than those of British and European populations but the differences are unlikely to be of practical significance. (The reader may wish to make an arbitrary deduction of, say 10 mm from each quoted figure to correct for this.)

Table 7.7. Grip reach envelope of seated person[a] (all dimensions are in millimetres).

Height above the SRP	Distance sideways								
	0	50	100	200	250	300	400	500	600
800	655	670	675	665	640	610	565		
700	695	710	720	715	705	695	670	625	
600	715	730	740	745	740	735	705	660	
550	720	740	750	760	760	750	720	675	610
500			750	765	765	760	730	680	615
450			745	770	770	765	735	680	620
400			735	770	775	770	735	675	620
350			725	765	775	770	735	665	615
300			710	755	765	770	725	755	605
200				735	755	755	705	610	585
100				690	715	720	660	550	530
0				635	670	680	615	470	470

[a]Horizontal forward reaches for a given height above the SRP and distance sideways from the centre line of the body. Figures tabulated are for a 5th %ile member of a population comprising of equal numbers of adult men and women.

7.3. *Zones of convenient reach*

At this point it is appropriate to develop the concept of a zone or space in which an object may be reached conveniently—that is without undue exertion. Consider what it means for a control to be 'within arm's length'. The upper limb, measured from the shoulder to the fingertip (or to the centre of grip), sweeps out a series of arcs centred upon the joint (see Figure 7.3). These define the zone of convenient reach (ZCR) for one hand, which extends sideways to the coronal plane of the body. The zones for the two limbs intersect in the midline (median) plane of the body. The volume which is thus defined comprises of two intersecting hemispheres. The radius of each hemisphere is the upper limb length (*a*) and their centres are a distance (*b*) equal to biacromial breadth apart.

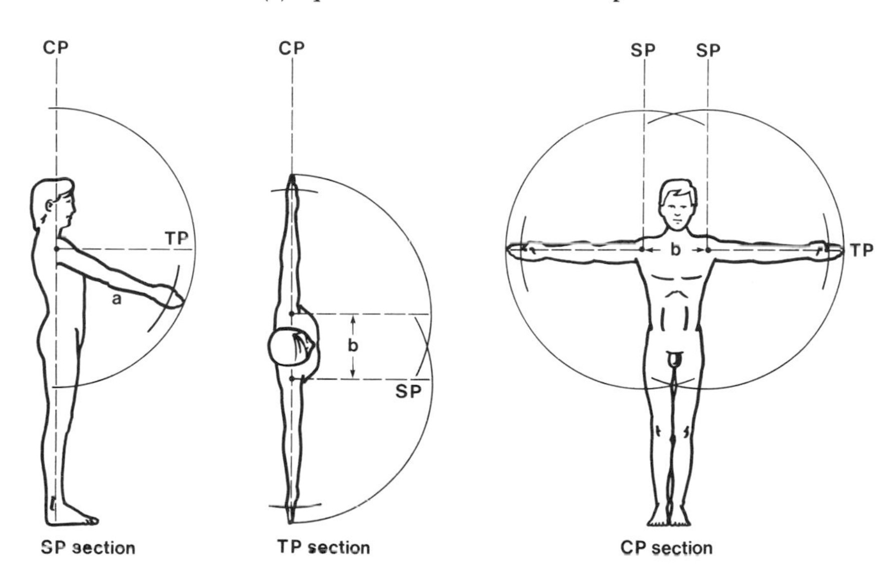

Figure 7.3. Zones of convenient reach (ZCR) seen in elevation and plan. (left to right) Vertical section in sagittal plane (SP) passing through shoulder joint; horizontal section in transverse plane (TP) passing through shoulder joints; vertical section in coronal plane (CP) passing through shoulder joints. Each plane of section is marked on the other two diagrams.

Many design problems are concerned with the intersection of vertical, horizontal or (very occasionally) oblique planes with either the volume of the workspace envelope or that of the zone of convenient reach. Suppose we wish to locate a set of items upon the vertical wall of a control room so that they might be conveniently operated by a standing person. The intersection of a plane with a sphere produces a circle. The radius of this circle may be calculated by Pythagoras' theorem as

$$r = \sqrt{a_2 - d_2} \tag{7.1}$$

where r is the radius of the circle on the wall, a is the upper limb length (or shoulder grip length) and d is the horizontal distance between the shoulder and the wall. Figure 7.4 shows the construction of such a zone for fingertip controls by a 95th %ile male or a 5th %ile female operator, assuming $d = 500$. The design also involves visual questions—optimal zones for visual displays have been added according to the criteria of Section 8.4.

The zone of convenient reach may be similarly described for any other vertical or horizontal plane parallel to a line joining the shoulders. Requisite data are given in Table 7.8.

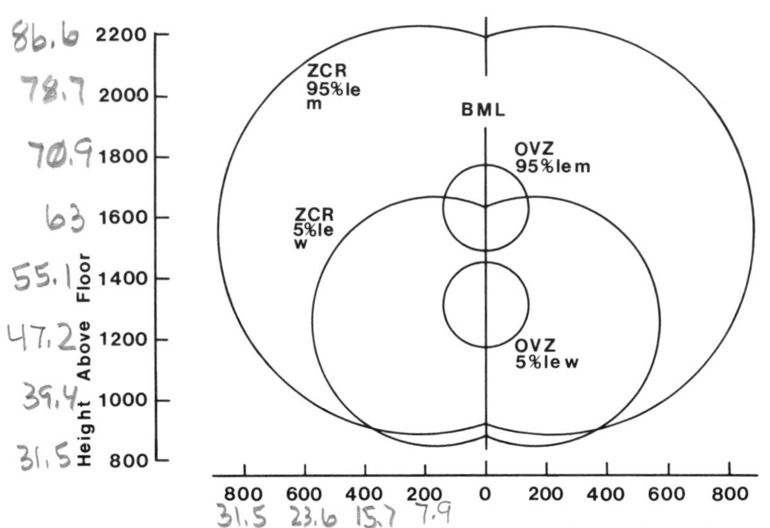

Figure 7.4. *Zones of convenient reach (ZCR) and optimal visual zones (OVZ) on a vertical surface 500 mm in front of the shoulders. Ninety-fifth %ile man (m) and 5th %ile woman (w). (BML = body mid-line.)*

7.4. *The 'normal working area'*

The intersection of a horizontal plane, such as a table or bench, with the zone of convenient reach defines what workstudy writers would call the maximum working area (Barnes 1958). Within this is a much smaller 'normal working area'—described by a comfortable sweeping movement of the upper limb, about the shoulder with the elbow flexed to 90° or a little less. Das and Grady (1983) have discussed this latter at length. The presentation which ensues (Figure 7.5) is based on the original concept of Squires (1956).

A person sits at a bench or table. His shoulder joints are located at S_1 and S_2 which are a distance b apart = biacromial breadth. The elbows are located at E_1 and E_2, at the table's edge, a distance d in front of the shoulders, such that d = abdominal depth/2. Both hands commence at H_1, the location of which is

Table 7.8. Zones of convenient reach[a] (all dimensions in millimetres).

	Radius (r)					
	Men			Women		
d	5th %ile	50th %ile	95th %ile	5th %ile	50th %ile	95th %ile
0	610	665	715	555	600	650
100	600	655	710	545	590	645
200	575	635	685	520	565	620
300	530	595	650	465	520	575
400	460	530	595	385	445	510
500	350	440	510	240	580	415
600	110	285	390			250

	Men[b]			Women[b]		
	5th %ile	50th %ile	95th %ile	5th %ile	50th %ile	95th %ile
Biacromial breadth	365	400	430	325	355	385
Shoulder height (standing, shod)	1340	1425	1560	1260	1335	1450
Shoulder height (sitting)	540	595	645	505	555	605

[a]To construct a zone of convenient reach in a vertical plane, a distance *d* in front of the shoulders, draw two circles of radius *r*; the centres of the circles are defined by standing or sitting shoulder height and biacromial breadth. To construct a zone of convenient reach in a horizontal plane, distance *d* above or below the shoulders, draw two semicircles of radius *r*, centred upon the position of the shoulders.
[b]Figures calculated from Equation 7.1 assuming a full grip.

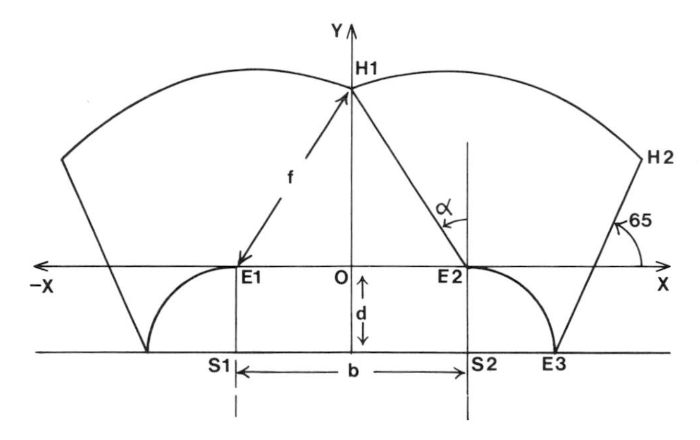

Figure 7.5. Construction of the normal work area (NWA).

calculated from the dimension f = elbow–grip length. Hence the angle a is given

$$\sin a = b/2f \tag{7.2}$$

When the elbow is flexed at 90° and the humerus is rotated at the shoulder about its own axis, the comfortable limit of outward rotation is limited to about 25°. In the present case the outer limit of the normal working area is a prolate epicycloid, H_1H_2, formed by two simultaneous rotations. The forearm (f) rotates through a + 25°, whilst the elbow itself moves outwards and backwards through a circular

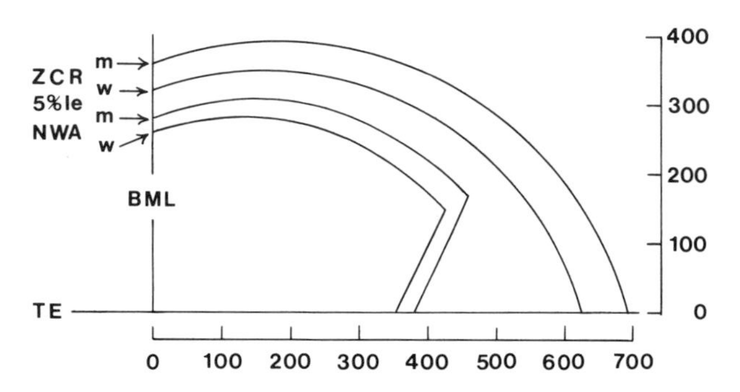

Figure 7.6. Zones of convenient reach (ZCR) and normal work area (NWA) on a table surface, for a 5th %ile man (m) and woman (w). (BML = body mid-line; TE = table edge.)

Table 7.9. Co-ordinates of the normal working area.[a]

		5th %ile man		5th %ile woman	
Position	Degrees	*X*	*Y*	*X*	*Y*
H_1	0	0	281	0	257
	10	56	298	53	271
	20	114	307	105	278
	30	172	307	160	279
	40	227	300	211	272
	50	281	287	260	258
	60	333	266	307	237
	70	370	239	350	211
	80	423	206	388	181
H_2	90	460	169	421	146
I	—	380	0	354	0

[a]Origin is at the table's edge in the midline of the body. The *X* axis runs along the table's edge; the *Y* axis is perpendicular. *I* is the point where the normal working area intersects the table's edge.

90° arc from E_1 to E_3. Hence, the forearm comes to lie at an angle of 90 − 25 = 65° to the table edge. Thus, when the arc *SE* has rotated through $\gamma°$, the arc *EH* has rotated through $\beta°$ such that

$$\gamma = \beta(a + 25)/90 \tag{7.4}$$

The co-ordinates of the elbow with respect to the shoulder are given by

$$X_1 = d \sin \beta \tag{7.4}$$

$$Y_1 = d \cos \beta \tag{7.5}$$

The co-ordinates of the hand with respect to the elbow are given by

$$X_2 = -f \sin (a - \gamma) \tag{7.6}$$

$$Y_2 = f \cos (a - \gamma) \tag{7.7}$$

Therefore the co-ordinates of the hand with respect to a point on the table's edge in the midline of the body are given by

$$X = d \sin \beta - f \sin (a - \gamma) + b/2 \tag{7.8}$$

$$Y = d \cos \beta + f \cos (a - \gamma) - d \tag{7.9}$$

Figure 7.6 and Table 7.9 are based on these equations together with the anthropometric data of Table 4.1.

7.5. *Joint ranges*

The flexibility of the human body is measured in terms of the angular range of motion of the joints. Joint movements are the subject of a terminology which is almost standardized (see Figure 7.7). Consider a vertical plane cutting the body down the midline into equal right and left halves—this is called the median (sagittal) plane. Any vertical plane parallel to it is called a sagittal plane and any vertical plane perpendicular to it is called a coronal plane. In general, sagittal plane movements of the trunk or limbs are called flexion and extension. (Flexion movements are those which fold the body into the curled-up foetal position.) Coronal plane movements are called abduction and adduction. (Abduction movements take a limb segment away from the midline.) Limb segments may also rotate about their own axes—inward (medially) or outward (laterally). Inward rotation of the forearm (turning the palm downwards) is called pronation; outward rotation (turning the palm upwards) is called supination.

There are surprisingly little joint range data available. Table 7.10 is based on a

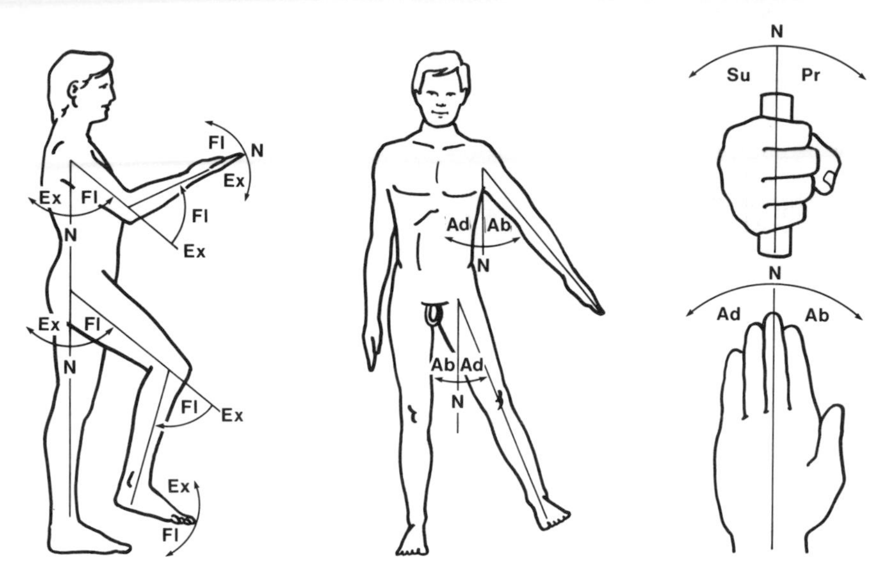

Figure 7.7. Terms used in the description of movements and joint ranges as given in Table 7.10. Fl = flexion; Ex = extension; Ab = abduction; Ad = adduction; Su = supination; Pr = pronation; N = neutral position.

survey of male US servicemen conducted by Dempster (1955), re-analysed by Barter *et al.* (1957) and quoted extensively (Damon *et al.* 1966, and elsewhere). Note that measurements were not necessarily made in the postures shown in Figure 7.7 (refer to Damon *et al.* (1966) for details).

In general, women have a somewhat greater flexibility than men (by about 5–15% on average). Decrements with age are probably small in the absence of joint disease (osteoarthritis, etc.) but since this is extremely common (universal in some populations) it is reasonable to assume greatly reduced flexibility in the elderly. Unfortunately, little statistical data are available.

The flexibility of one joint may be influenced by the posture of adjacent joints—the most important example is flexion of the hip which is very much greater when the knee is flexed than when it is extended. (Prove this by touching your toes.)

Table 7.10. Joint ranges (degrees).

Joint		5th %ile	50th %ile	95th %ile	SD
1.	Shoulder flexion	168	188	208	12
2.	Shoulder extension	38	61	84	14
3.	Shoulder abduction[a]	106	134	162	17
4.	Shoulder adduction	33	48	63	9
5.	Shoulder medial rotation	61	97	133	22
6.	Shoulder lateral rotation	13	34	55	13
7.	Elbow flexion	126	142	159	10
8.	Pronation[b]	37	77	117	24
9.	Supination[c]	77	113	149	22
10.	Wrist flexion	70	90	110	12
11.	Wrist extension	78	99	120	13
12.	Wrist abduction (radial deviation)	12	27	42	9
13.	Wrist adduction (ulnar deviation)	35	47	59	7
14.	Hip flexion[d]	92	113	134	13
15.	Hip abduction	33	53	73	12
16.	Hip adduction	11	31	51	12
17.	Knee flexion	109	125	142	10
18.	Ankle flexion (plantar flexion)	18	38	58	12
19.	Ankle extension (dorsiflexion)	23	35	47	7

[a] Accessory movements of spine increase this to 180°.
[b] Rotation of the forearm about its own axis such that the palm faces downwards.
[c] Rotation of the forearm about its own axis such that the palm faces upwards.
[d] Measured with the knee fully flexed. If the knee is extended the range will be very much less (approx. 60°).
Data from Barter *et al.* (1975).

Chapter 8
Posture

The posture which a person adopts when performing a particular task is determined by the relationship between the dimensions of the person's body and the dimensions of the various items in his or her workspace (a tall person using a standard kitchen will stoop more than a short one, etc.). The extent to which posture is constrained in this way is dependent upon the number and nature of the connections between the person and the workspace. These connections may be either physical (seat, worktop, etc.) or visual (location of displays, etc.). If the dimensional match is inappropriate the short- and long-term consequences for the well being of the person may be severe. The purpose of this chapter is, firstly, to explore the consequences of poor posture and, secondly, to enumerate ways in which this unsatisfactory state of affairs may be avoided. Hence, we shall develop postural criteria for the application of anthropometrics to workspace design.

8.1. The physiology and pathology of posture

Posture may be defined as the relative orientation of the parts of the body in space. To maintain such an orientation over a period of time, muscles must be used to counteract any external forces acting upon the body (or in some minority of cases internal tensions within the body). The most ubiquitous of these external forces is gravity. We shall not embark at this point upon a detailed biomechanical analysis of posture—the interested reader is referred to Grieve and Pheasant (1982). Instead, let us take a simple illustrative example. Consider a standing person who leans forwards from the waist. The postural loadings on the hip extensor or the back extensor muscles are proportional to the horizontal distance between the hip and lumbosacral joints, respectively, and the centre of gravity of the upper part of the body (i.e., the head, arms and trunk). The farther the trunk is inclined the greater this distance becomes (Figure 8.1). Some physiologists have chosen to call the muscular activity which results from this loading 'postural work' or 'static work'. To the physicist the latter is a contradiction in terms (since by definition work is performed when a force moves its point of application). To avoid this difficulty we shall use the term 'postural stress'—noting that our usage conforms with the overall scheme for the description and analysis of workload advocated by the late J. S.

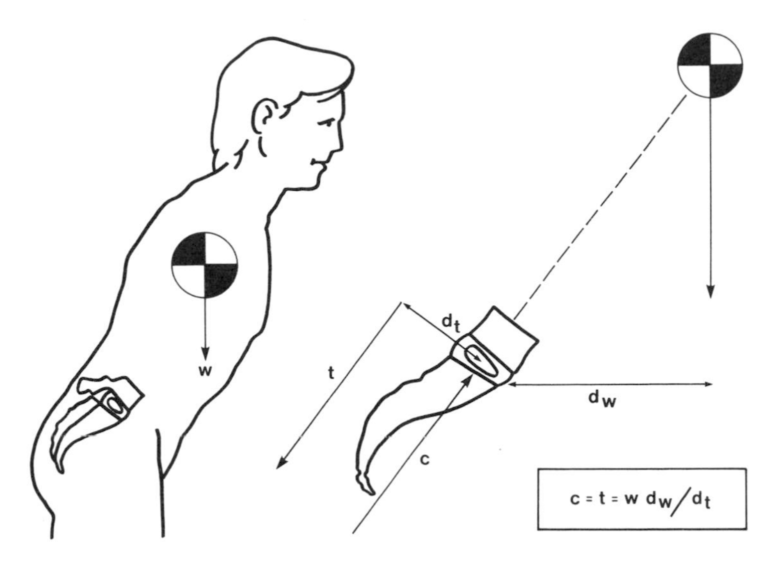

Figure 8.1. Biomechanical analysis of postural stress in a forward leaning position. Note that this analysis ignores the direct effect of the weight of the trunk which the spine must support even when $d_w = 0$.
w *is weight of that part of the body above the lumbo-sacral joint;* c *is the compressive force acting along the axis of the spine;* t *is the tension in the back muscles (erector spinae).*

Weiner (1982). In general, it will be argued, the best working posture is the one which imposes least postural stress.

If postural stress is prolonged (for more than a few minutes) there are certain untoward consequences which, again following Weiner (1982), we may characterize as 'postural strain'. Muscle as a tissue responds badly to prolonged static mechanical loading. (The same is probably true of other soft tissues, and even perhaps of bone, but the physiology of these cases is much less well understood.) Static effort restricts the flow of blood to the muscle. The chemical balance within the muscle is disturbed, metabolic waste products accumulate and the condition of 'muscular fatigue' supervenes. The person experiences a discomfort which is at first vague but which subsequently develops into a nagging pain until it becomes a matter of some urgency that relief is sought by a change of position. Should you require evidence of this course of events, you should raise one of your arms and hold it out in front of you as you continue to read (or attempt to do so). Provided our workspace and/or working schedule allows us to make the frequent shifts of posture which are subjectively desirable, all will be well—since the physiological processes of muscular fatigue are relatively rapidly reversible by rest or change of activity (particularly if the activity involves stretching the fatigued muscle).

In general, we may think of 'fidgeting' as our bodies' defence against postural stress. This mechanism characteristically operates at a subconscious level—usually we fidget before we become consciously aware of discomfort. In relaxed sitting the sensory stimuli probably come more from the compression of the soft tissues of the buttocks and thighs than from muscle tension. The crossing and uncrossing of the legs is a characteristic way of redistributing the pressure on the buttocks and, hence, pumping blood through the tissues. The rate of fidgeting can be used as an index of the comfort of chairs—the less comfortable we are, the more we fidget. It is a matter of common experience that other factors are involved. Some people fidget more than others and we all fidget more when we are bored—presumably because mental activity can 'shut out' the sensory stimuli which cause the fidgeting (or raise our threshold of discomfort). Such a hypothesis is in line with contemporary theories of the nature of pain (Melzack and Wall 1982). Students almost universally consider lecture theatre seating to be uncomfortable—is this to do with the seats or the lectures?

Physiologically, comfort is the absence of discomfort—I know of no nerve endings capable of transmitting a positive sensation of comfort from a chair. Comfort is a state of mind which results from the absence of unpleasant bodily sensations. (The same relationship does not hold, however, for pleasure and pain.)

Suppose that the working circumstances are such as to closely constrain us to a particular posture and prevent postural change—the consequences may be divided into those occurring over the short term and those occurring over the long term. In the short term, mounting discomfort may distract the operator from his task leading to an increased error rate, reduced output, accidents, etc. From the physiological standpoint, however, we are still talking about reversible strain—since the symptoms are relieved by rest or by a change of activity. At some point, nevertheless (and this point is not well defined since the transition is probably gradual rather than sharp), pathological changes in the muscle or soft tissue take over. Typically, pain comes on after increasingly short periods of postural stress and rest is less certain to bring relief. At this point we are dealing not with discomfort but with disease.

A number of painful afflictions of the musculoskeletal system are considered to be associated with posture. The lumbar region, the neck, the shoulder and the forearm are the parts of the body most commonly affected—probably in that order. Both clinicians and lay people have a plethora of descriptive names for these conditions. Unfortunately the terms overlap and synonyms abound, often reflecting a lack of real agreement as to the underlying causes of the symptoms. It is probable that the proximate cause in many cases is a cyclic process of pain and muscle spasm together with the inflammation of surrounding soft tissue. Whether the inflammation causes the spasm or vice versa is unclear, as is the extent to which postural stress actually initiates the symptoms rather than merely aggravating pre-existing damage.

In some circumstances the inflammation may be relatively localized around tendinous attachments of the muscles (particularly in the shoulder, elbow or forearm) and in these cases the effects of posture may be indirect ones. For example, inflammation of the tendons of the forearm (tenosynovitis) is very common in people who perform certain industrial assembly tasks. It is generally considered that this disorder is caused by repetitive motions of the forearm and wrist and the cumulative wear and tear which results. There is little doubt, however, that the posture of the wrist may greatly increase or diminish the damage which occurs (see Section 8.6).

Muscular tension, particularly in the neck, shoulders and back, may also be associated with psychological stress and, since worktasks performed in poor postures may also be frustrating in other respects, the effects may be interrelated. (Repetitive data-entry tasks at VDU terminals are a case in point.) Some people are more 'tense' than others in these situations and personality differences may be of some importance in this respect. Indeed, even in the short term, people vary greatly in their tolerance of postural stress. A poorly designed seat or awkward position which is only mildly uncomfortable for one person may rapidly become agonizing for another. It is unclear whether this is due to differences in musculoskeletal fitness or to more obscure differences within the central nervous system.

The spine is particularly susceptible to postural stress. The long muscles of the back, known as the erector spinae group, are important in lifting actions and in supporting the weight of the upper part of the body if the trunk is inclined forwards. Since they run more or less parallel to the vertebral column, any tension in these muscles will exert an equal and opposite compression on the spine (see Figure 8.1). Between each adjacent pair of vertebrae lies an intervertebral disc—a small pad of fibrocartilage (i.e., gristle), which acts as an hydraulic shock absorber. The discs have two parts, known to anatomists as the annulus fibrosus and the nucleus pulposus (Latin terms which, if roughly translated into English as the 'fibrous outside' and 'squidgy middle', adequately describe the properties of the structures concerned) (Figure 8.2). By virtue of its chemical consistency, the nucleus tends to imbibe (absorb) water from its surroundings and in so doing

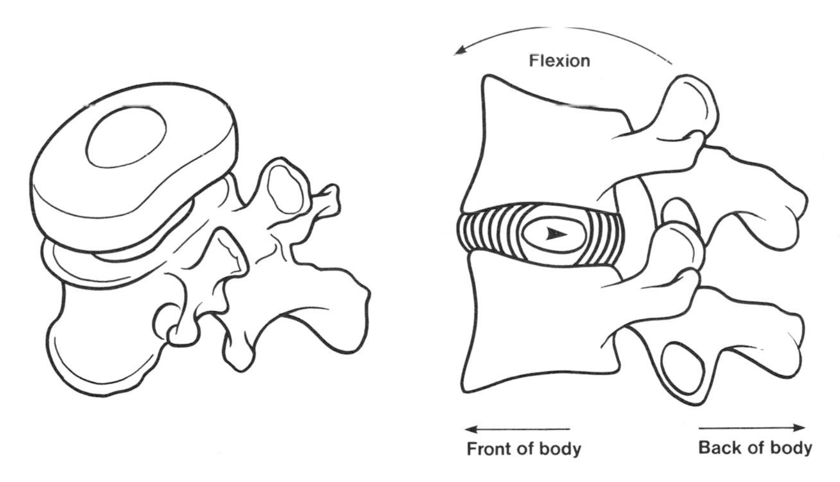

Figure 8.2. (left) Lumbar vertebra surmounted by intervertebral disc, showing outer and inner parts. (right) Deformation of the disc during flexion of the spine.
Redrawn from Kapanji (1974).

swells, stretching the annulus until an equilibrium condition is reached. (Imagine a self-inflating motor-car tyre.) The equilibrium condition is also affected by the level of compressive loading to which the spine may be subjected. During sleep the spine is unloaded and water flows into the disc, but if we stand or sit up straight the spine is automatically loaded (at the lumbosacral level by approximately one-half of the body weight). The postural activity of the erector spinae, if we stand or sit in a forward leaning position, will tend to drive water from the disc. Hence, we are a little shorter at the end of our working day than at the beginning.

In the standing position the spine presents a concavity (lordosis) in the lumbar region. As we lean forwards part of the motion occurs at the hips but part occurs by flexion of the lumbar spine, i.e., by first a flattening and finally a reversal of the lumbar concavity. For this to occur the disc must be deformed into a wedge shape—the nucleus is driven backwards and the posterior part of the annulus comes under tension (Figure 8.2). Lifting weights in this position is particularly hazardous. If the loading is sufficiently great the annulus will rupture and the nucleus will be extruded under pressure. This is known medically as a prolapsed or herniated intervertebral disc, and to the lay person as a 'slipped disc'. Extruded disc fragments may press upon nerves leading to an agonizing (and totally incapacitating) pain in the back ('lumbago') which may also radiate down the lower limb ('sciatica').

Although both patients and their physicians commonly attribute acute back pain attacks to lifting actions, or to a variety of other 'causes' such as twisting, falling or even coughing and sneezing, such an attribution is probably incorrect in most cases. It is much more probable that the disc herniation is the endpoint of a long process of cumulative damage and degeneration. Some authorities argue that disc degeneration is an inevitable part of the ageing process—and it is certainly true that few of us are destined to go through life with a spine which is structurally and functionally as sound as it was in our youth. A variety of factors may accellerate disc degeneration leading to its appearance in relatively young people. Postural stress, both from forward leaning positions and from prolonged sitting (see Section 8.7), is generally held to be an important causative factor in disc degeneration (as indeed are lifting and heavy exertion, vibration and the repetitive jolts and impacts which occur in motor vehicles), but there is little general agreement amongst clinicians as to the percentage of serious back problems in which a disc lesion *per se* is involved.

A posture which is adopted to perform a certain activity may, in the long term, first become habitual (outside the context of the activity) and finally become irreversible. A number of occupational groups, from lathe operators and bricklayers to dentists and draughtsmen, have a propensity to develop spinal deformities associated with the stooped or twisted positions in which they work (but it seems that these do not necessarily give rise to painful symptoms). Curiously enough, although these deformities are a matter of common experience, and were remarked upon as long ago as 1713 (Ramazzini 1713), they have received remarkably little scientific investigation. Some authorities consider that their consequences for general health may be severe—since the functioning of a variety of internal organs may be impaired. Such assertions, although based on plausible anatomical and physiological reasoning, must be regarded as unproven.

Among the many authors who have discussed the applied physiology of working posture are Floyd and Ward (1966), Van Wely (1970), Corlett and Manenica (1980), Corlett (1983), Grieve and Pheasant (1982) and Chaffin and Andersson (1984). Wickstrom (1979) and Andersson (1979) have given balanced overviews of back pain in the occupational setting. Occupational disorders of the neck, shoulder and arms have been reviewed by Waris (1979) and Maeda (1977). Authoritative discussions of musculoskeletal disorders in general include Cyriax (1978), Caillet (1977) and Travell and Simons (1983). The postural afflictions of various occupational groups have been reported *inter alia* by Duncan and Ferguson (1974), Grandjean and Hünting (1977), Westgaard and Aaras (1980), Hünting *et al.* (1980, 1981), Vihma *et al.* (1982), Grandjean *et al.* (1983, 1984) and Kukkonen *et al.* (1983).

I am well aware that the preceding account is an oversimplification of what in actuality is a very complex problem.

8.2. Sources of postural stress in workspace design

The consequences of prolonged static muscular loading may be severe. How may this loading be avoided? There follows a list of simple guidelines based in part upon Corlett (1983). If these were generally followed, particularly in the design of industrial workstations, a dramatic reduction in the incidence of musculoskeletal disease could be anticipated. Case studies showing that such a reduction can in fact be achieved by ergonomic intervention include Westgaard and Aaras (1980), Kukkonen *et al.* (1983) and Grandjean *et al.* (1984).

(i) Encourage frequent changes of posture

Sedentary workers, therefore, should be able to sit in a variety of positions—some office chairs are now being designed with this in mind. For many industrial tasks a sit–stand workstation is to be advocated. The task is typically set at a height which is suitable for a standing person (see Section 8.5) and a high stool or 'perch' is provided as an alternative. There seems little doubt that most sedentary workers would be better off if their jobs required them to get up and move around once in a while. Perhaps we should deliberately design a moderate degree of physical activity (particularly activity involving stretching) into work tasks. This approach should be treated with caution since we would not wish to overstrain stiff sedentary bodies—none the less, the idea is certainly worth further consideration. The introduction of periods of exercise into the working day is beginning to find favour, particularly in Scandinavia.

(ii) Avoid forward inclination of the head and trunk

This commonly results from visual tasks, machine controls or working surfaces which are too low (see Sections 8.4 and 8.5).

(iii) Avoid causing the upper limbs to be held in a raised position

This commonly results from a working level which is too high (or a seat which is too low). If manipulative tasks must be performed in a raised position, perhaps for visual reasons, arm supports should be provided. In addition to being a considerable stress to the shoulder muscles, tasks which must be performed at above the level of the heart impose an additional circulatory burden. The upper limit for manipulative tasks should be around halfway between elbow and shoulder level.

(iv) Avoid twisted and asymmetrical positions

These commonly result from expecting an operator to have eyes in the back of his head, i.e., from the mislocation of displays and controls.

(v) Avoid postures which require a joint to be used for long periods at the limit of its range of motion

This is particularly important for the forearm and wrist.

(vi) Provide adequate back support in all seats

It may be that for operational reasons the backrest cannot be used during the performance of the work task—but it will still be important in the rest pauses.

(vii) Where muscular force must be exerted the limbs should be in a position of greatest strength

Unless by so doing one of the foregoing rules is broken (see Section 8.3).

(viii) When the weight of the body must be supported avoid the build up of pressure on sensitive areas of tissue

8.3. Posture and strength

The force which an individual is capable of exerting in any kind of physical action is almost always strongly influenced by the posture which he or she adopts. For practical design purposes questions of strength are almost always questions of posture as well—so the two topics will be discussed together. Studies in which strength is measured in a number of different working positions generally show that the differences between conditions (i.e., between postures) are substantially greater than the differences between people in any one condition (see, for example, Pheasant *et al.* 1982).

The term 'strength' will, for present purposes, be defined as meaning the maximum steady force or torque which an individual can exert in a static exertion of short duration (e.g., 3–5 s). Exertions of longer duration are properly described as measures of endurance. Measures of power output can legitimately be given the name 'dynamic strength' (but the increasingly common application of this latter term to subjectively determined acceptable workloads is confusing and should be avoided). Sources of information concerning static strength are numerous; studies of dynamic strength are rare. The simple reason for this is that it is easy to build a test rig to measure the forces subjects can exert against an immovable resistance, but achieving a controlled resistance under dynamic conditions is much more complicated. (Some progress is being made in this area (Grieve 1984).) In practice, as we shall see, the limitations of static strength data are not as great as they may seem.

Designers and other practical people commonly ask three types of question concerning human strength. The differences are subtle but important:

Question A. What forces can people exert under one set of circumstances relative to those they can exert under another (e.g., what posture or control location is 'best' for a particular action)?

Question B. What actual forces are the members of a particular target population capable of exerting under particular conditions?

Question C. What forces can the members of a population be reasonably expected to exert in a particular activity according to certain criteria of operational relevance (e.g., productivity, convenience, safety)?

Question A is probably the most common and, fortunately, it is the easiest to answer—from first principles, from published data or by a relatively simple experiment. Question B is more difficult since strength is highly population-specific and modelling or estimation techniques in this area are in their infancy. (We should note that the variation in strength both within and between population groups is probably greater than that of linear dimensions.) In any case, we should never attempt Question B until Question A has been answered—and then it can sometimes be rephrased with advantage, e.g., is it possible for the 5th %ile person to exert x units of force in this particular operation? Question C is commonly very difficult to answer and sometimes impossible. Again, this question may sometimes be rephrased with advantage, e.g., how may we make this lifting and handling task safer? (See Section 14.2.)

There are several different interlocking mechanisms by which posture affects strength. Consider the function of the elbow flexor muscles (biceps brachii, brachialis, etc.) in the pulling actions shown in Figure 8.3. As a joint changes its position (for example from the left-hand to the right-hand position) muscles which cross the joint change in length and, hence, in their capacity to exert tension (in general, the latter is greatest when the muscle is at its greatest length), and also in their leverage about the joint centre. The combined effects of these changes determine the shape of an angle–torque curve which describes the flexor or extensor moment that the muscles acting about the joint can exert at any given joint angle—some examples of this curve are given in Figure 8.4. Note that in each case the action is strongest in a position at or close to the one in which the muscles are at their greatest length. Since many muscles cross more than one joint, the

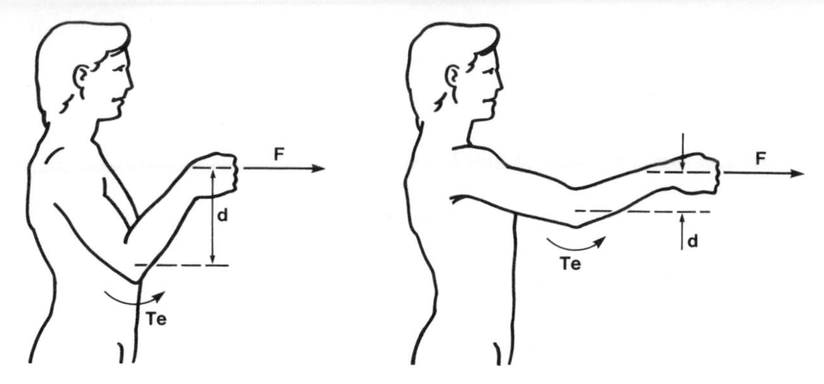

Figure 8.3. Torque about the elbow (T_e) *required to exert the same pulling force* (F) *in two different postures.*

torque available at each may also be dependent upon the angles of neighbouring joints—but few of these interactions have been worked out at the present time (Grieve and Pheasant 1982).

Let us now return to Figure 8.3. Suppose the person exerts a force of F kg in each of the postures shown. The exertion demanded from the elbow extensors will be different in each posture and will be given by the equation

$$T_e = Fd \tag{8.1}$$

where d is the perpendicular distance from the line of action of the force to the elbow joint. In this case we are ignoring the weight of the arm. We could calculate

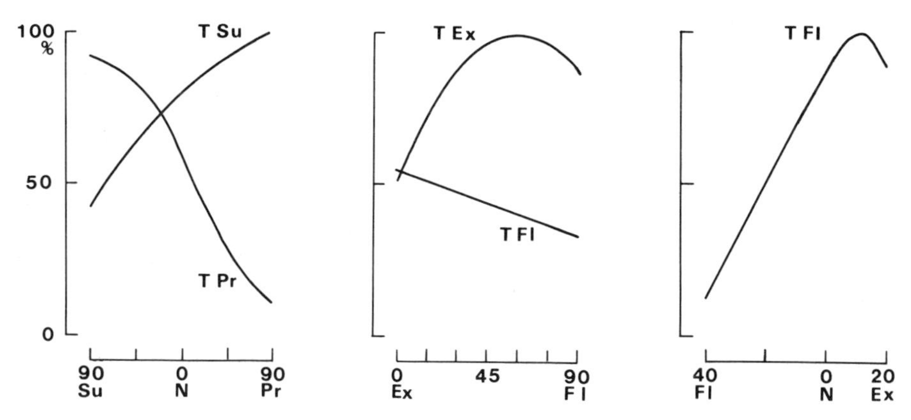

Figure 8.4. Angle–torque curves: (left) supination (T_{Su}) *and pronation* T_{Pr}) *of the forearm (Grieve and Pheasant 1982); (centre) flexion* (T_{Fl}) *and extension* (T_{Ex}) *of the knee (Smidt 1973); (right) flexion* (T_{Fl}) *of the ankle (Grieve and Pheasant 1982). In each case the torque is plotted as a percentage of the maximum torque found in any position against the angular position of the joint as defined in Figure 7.7. Smooth curves were drawn by eye in the centre and right-hand figures.*

the loading on any other joint in the same way. The complex mathematical models which have been used to predict human strength (Chaffin and Andersson 1984) work by performing this calculation for each joint of the body in turn and asking whether the torque required exceeds the capacity of the muscle group.

There are circumstances in which the force a person can exert is determined by factors other than the strength of his or her muscles as such. In certain lifting, pushing or pulling actions body weight and equilibrium, or the frictional purchase between the feet and the floor, may be limiting factors. To some extent a subject may trade-off one factor against another, and the adoption of an optimal strategy in this respect may be considered to be a characteristic of skilled performance. This subject (which is not really central to our present concerns) is considered at length in Grieve and Pheasant (1982).

What percentage of an individual's maximal strength is it right and proper to demand in a particular activity? The answer will depend on how critical the activity is, how frequently it is to be performed and many other factors. Certain rules of thumb exist but these are little more than figures 'plucked out of the air'. One could suggest that 15% is the maximum allowable for continuously operated controls, 30% for frequently operated controls and 60% for occasional exertions. Anything above 60% should be avoided—unless, of course, you are interested in weight training the operator!

8.4. *Vision and the posture of the head and neck*

The visual demands of a task and the location of visual displays are important not only in themselves but also because they largely determine the posture of the head and neck. Look carefully at the printed text on this page—fix your eyes on one particular word near the centre of the page. You will find that other words become less distinct with increasing distance from the central point of fixation and the margins of the page are no more than an indistinct blur. Only the central part of the visual field is sufficiently sensitive for demanding visual tasks such as reading text or recognising a face. The area of foveal vision, as this central region is called, is limited to a solid angle of some 5° about the line of central fixation. Visual work demands that the foveal regions of both eyes be directed convergently upon the task. Furthermore, the lenses of the eyes must accommodate (focus) to the appropriate distance. The processes of direction and convergence of gaze are integrated with accommodation by a set of reflexes so finely tuned that we are unaware of their existence until such times as they break down by reason of age or misuse.

If we sit or stand with our head up, and look ahead, our eyes will naturally assume a slight downward gaze of some 10 or 15° from the vertical—this we shall call the relaxed line of sight. The direction of gaze is altered, firstly, by movements of the eyeballs in their sockets (orbits) by means of the orbital muscles and, secondly, by movements of the head and neck. Taylor (1973) states that the eyes may be raised by 48° and lowered by 66° without head movements. In practice, only a part of this range of movement is used. Weston (1953) in his classic study of visual fatigue suggests that, in practice, downward eye movements were limited to 24–27°; beyond that point the head and neck are inclined forwards and the neck muscles come under tension to support the weight of the head (see Figure 8.5).

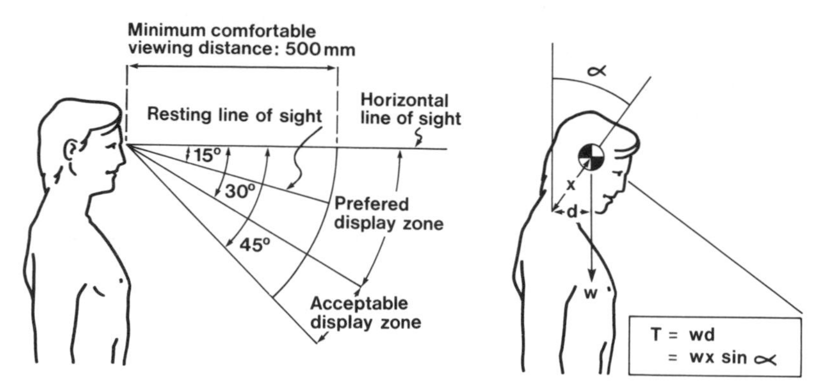

Figure 8.5. (left) Preferred viewing conditions as described in text. (right) Postural stress to neck muscles resulting from a downward line of sight.
T *is the torque about the neck;* w *is the weight of the head and neck;* x *is the distance from C7 to the centre of gravity of the head and neck.*

Grandjean *et al.* (1984) described an experiment in which a group of VDU operators were given an adjustable workstation and encouraged to set it to their own satisfaction over a period of 1 week—the preferred visual angle was 9° [4·5°] downwards from the horizontal. Brown and Schaum (1980) have also conducted fitting trials on VDU workstations. Their results are reported in co-ordinate form but it is possible to calculate that the average preferred visual angle was 18° downwards.

On the basis of the above findings we may conclude that the preferred zone for the location of visual displays extends from the horizontal line of sight downwards to an angle of 30° and that the optimal line of sight is somewhere in the middle of this zone. Given that some modest degree of neck flexion is acceptable this could be extended a further 15° (see Figure 8.5). Visual comfort and satisfactory posture is also dependent upon displays being located a suitable distance from the eyes. When focused on infinity, or any object more than around 6 m distant, the lens of the eye is completely relaxed. To look at closer objects than this requires effort, both of the orbital muscles for convergence and of muscle within the eye itself for accommodation. In young people the processes of convergence and accommodation reach their limits or 'near points' at around 80 and 120 mm, respectively. (The latter increases dramatically with age as the lens of the eye stiffens—this also reduces the rate at which the eye can accommodate to different distances). Visual work performed excessively close to the eyes is fatiguing and leads to 'eyestrain'—a poorly defined condition involving blurring of vision, headache and burning or 'gravelly' sensations around the eyes. As is the case with most criteria there is no sharp cut-off point for minimum acceptable viewing distance and authorities differ in the figures they recommend. For most practical purposes we could consider 350 mm an absolute minimum but 500 mm is safer and as much as 700 mm may be desirable. The VDU operators studied by Grandjean *et al.* (1984) adjusted their workstations to an average visual distance of 760 mm (settings ranged from 610 to 930 mm). The data of Brown and Schaum (1980) give an average preferred figure of 624 mm (see also pp. 193–197).

It is interesting to note that pain and spasm in the neck muscles (trapezius, sternomastoid, splenius, etc.) can lead to 'mechanical headache'—experienced in various parts of the head and face and not uncommonly around or 'behind' the eyes (Travell 1967, Dalassio 1980, Travell and Simons 1983). (Anatomists reading this will note the proximity of the proprioceptive supply of these muscles to the spinal nucleus of the trigeminal nerve.) The symptoms of mechanical headache and eyestrain are exceedingly similar.

Head and neck posture in the workplace can commonly be improved by the provision of raised reading stands. For VDU operators these may be located next to the screen but for typists they are better behind the typewriter. The benefits to be gained from these simple devices can be considerable. (I have seen overnight improvements in the neck and shoulder symptoms of typists result from the provisions of such a stand.)

8.5. *Working height*

The height above the ground at which manual activities are performed by the standing person is a major determinant of that person's posture. If the working level is too high the shoulders and upper limbs will be raised, leading to fatigue and strain in the muscles of the shoulder region (trapezius, deltoid, levator scapulae, etc.). If any downward force is required in the task the upper limbs will be in a position of poor mechanical advantage for providing it. This problem may be avoided if the working level is lower. One commonly hears people talk of "using their weight" or "getting their weight on top of" the action. This is probably a misconception—what we really mean is that a vertical force may be exerted with minimal loading to the elbow and shoulder extensor muscles. A downward force, however exerted, can never exceed body weight (unless your feet are bolted to the floor), but in some positions the muscles of your arm may lack the strength to lift your feet off the ground.

If, however, the working level is too low the trunk, neck and head will be inclined forwards with consequent postural stress for the spine and its muscles. It may be presumed that somewhere between a working level that is too high and one which is too low there may be found a suitable compromise at which neither the shoulders nor the back are subjected to excessive postural stress.

It is important to distinguish between working height and worksurface height. The former may be substantially higher than the latter if hand tools or other equipment are being used in the task. In some cases the working level may actually be below the worksurface—consider the task of washing up which, in the conventional kitchen, is performed in a recess set into the working surface (i.e., the sink).

Grandjean (1981) summarises recommendations for working height as follows (see also Sections 2.4 and 15.1):

(*i*) 50–100 mm above elbow level for delicate or precise work;
(*ii*) 50–100 mm below elbow level for general light manipulative tasks;
(*iii*) 150–400 mm below elbow level for heavy work, particularly involving downward forces

8.6. *The forearm*

In general, manipulative tasks are best performed with the wrist in a neutral position—in which the middle finger is aligned with the axis of the forearm (see Figure 8.6). Departure from this alignment, in the performance of manipulative tasks, is associated with pain and inflammation in the forearm muscles and associated structures. The commonest problems arise from a position of adduction (ulnar deviation—see Figure 7.7). This may occur in a keyboard operation—often because the keyboard is too high, leading to an outward movement of the elbows (i.e., by shoulder abduction) and a compensatory deviation of the wrist. Kroemer (1972) argues that the horizontal position of the palm which the conventional keyboard demands is undesirable—if the operator is to achieve this position and still keep the upper arms hanging vertically downwards from the shoulders, an extreme pronation of the forearm is required. Since this extreme position is itself fatiguing the shoulders will be abducted. Kroemer's proposed solution, based on sound biomechanical principles, is to separate the right- and left-hand groups of

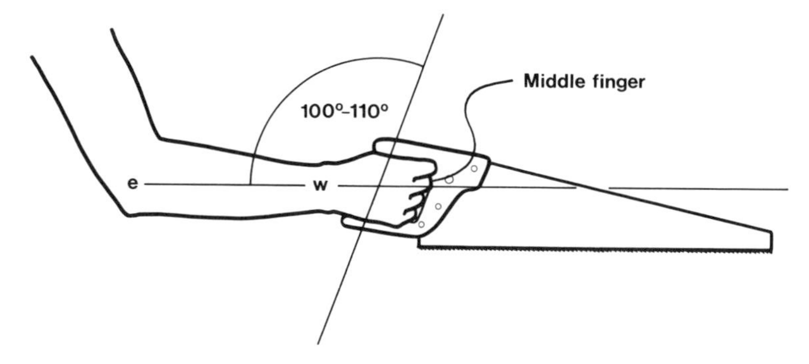

Figure 8.6. The neutral position of the wrist is preserved if the axis of grip makes an angle of 100–110° with the axis of the forearm.

keys and place them on sloping surfaces clustered about convergent axes with a wedge-shaped space between.

Ulnar deviation may also be consequent upon the design of hand tools. The axis of a cylindrical type handle forms an angle of 100–110° with the axis of the forearm when the wrist is in a neutral position (Barter *et al.* 1957). It is interesting to note that many traditional tools have evolved forms which preserve this orientation. Adduction of the wrist is particularly undesirable in tasks which require repetitive rotational movements of the forearm (pronation and supination) since the range of these movements is reduced. This means that the number of movements required to perform a certain task will be increased with a consequent increased probability of overstrain. Tichauer (1978) showed that the provision of pliers designed to maintain the neutral position of the wrist reduces the incidence of forearm ailments in assembly workers (see Figure 8.7).

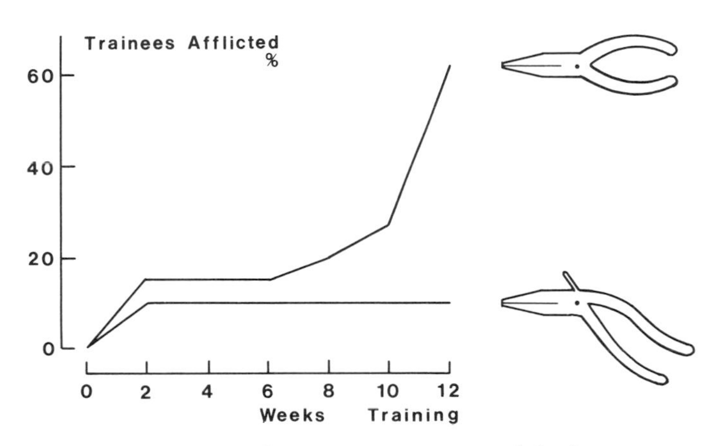

Figure 8.7. Incidence of repetitive strain injuries of the forearm in groups of trainee electronics assembly workers using conventional pliers (upper) and re-designed pliers (lower). Figures for tenosynovitis, carpal tunnel syndrome and epicondylitis (tennis elbow) have been lumped together. Data from Tichauer (1978).

8.7. Sitting posture

Consider what happens when you sit down on a relatively high seat (such as a dining-room chair). You flex your knees through 90° and make another 90° angle between your thighs and trunk. Most of your weight is taken by the ischial tuberosities—two bony prominences which you can feel within the soft tissue of your buttocks if you sit on your hands. Part of the right angle between the thighs and trunk is achieved by flexion at the hip joint. After an angle of 60° is reached this movement is opposed, unless we are very flexible, by tension in the hamstring muscles (located in the backs of the thighs); hence we tend to complete the movement by a 30° backward rotation of the pelvis (see Figure 8.8).

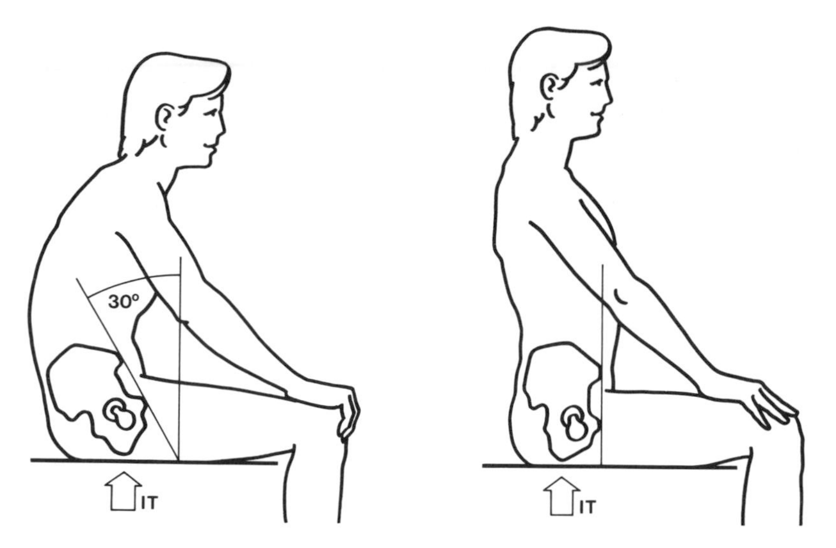

Figure 8.8. In relaxed sitting (left) the pelvis rotates backwards and the spine is flexed. To sit up straight (right) requires muscular exertion to pull the pelvis forward. The ischial tuberosities (IT) act as a fulcrum.

This must be compensated by an equivalent flexion of the lumbar spine, if the overall line of the trunk is to remain vertical. Hence, in sitting down we tend to lose the concavity (lordosis) which is characteristic of the lumbar region of a well-formed human spine in the standing position. This then is the fairly typical relaxed or slumped sitting position—which is generally considered to be a bad thing. In order to 'sit up straight' and regain our lost lordosis we must make a muscular effort to overcome the tension in the hamstrings. (The effort probably comes from a muscle deep within the pelvis called iliopsoas.) We cannot merely relax the hamstrings since their tension is a passive one; caused by the stretching of tissue (just like an elastic band) rather than by actual muscular contraction.

A major objective in the ergonomic design of seating should be to maintain a modest degree of lordosis without the need for muscular tension—hence allowing the user to adopt a posture which is both physiologically satisfactory for the spine and comfortably relaxed. Conventionally, this is achieved by:

(*i*) an obtuse angle (around 110°) between seat and backrest;
(*ii*) a seat which is neither lower nor deeper than necessary;
(*iii*) a backrest which is contoured to the desirable shape of the spine (see Chapter 11).

The well-designed backrest will not only help maintain the spine in a desirable postural configuration but it will also support part of the weight of the trunk and, hence, directly reduce the mechanical loading on the spine. This latter effect increases with the rake angle of the backrest. (This may be predicted theoretically and has been experimentally demonstrated by Andersson *et al.* 1974.) It will tend to be lost in tasks such as writing or typing which require us to lean forwards but the backrest remains important in these activities, for support during rest pauses. Grandjean (1981) describes a study of office workers using time-lapse photography, which showed them to be in contact with the backrest for 42% of the time.

In recent years a radical new approach to seating has been proposed, Mandal (1976, 1981) has argued (quite cogently in my view) that seat surfaces should slope forwards, hence diminishing the need for hip flexion (particularly in tasks such as typing and writing) and encouraging lumbar lordosis. A number of seat designs now incorporate a tilt mechanism (see Figure 8.9). The disadvantage of such a

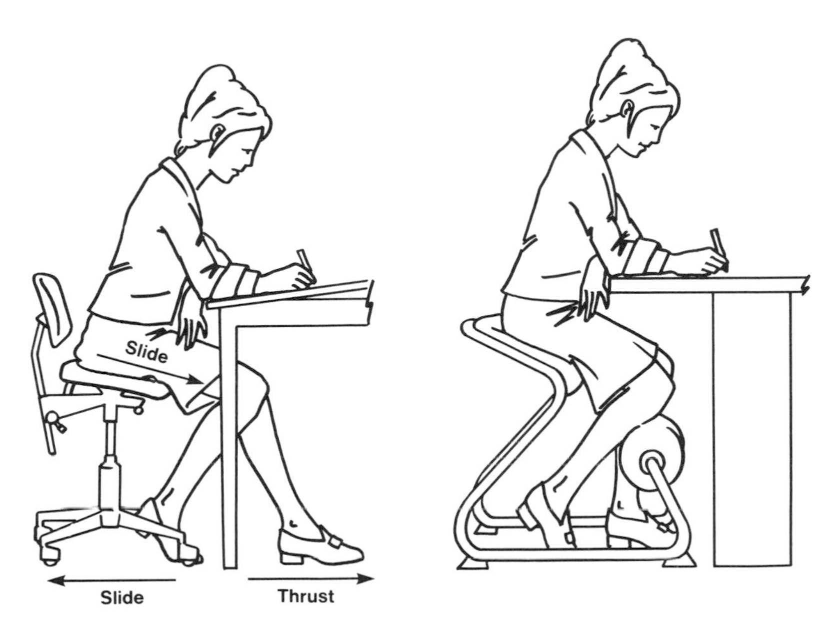

Figure 8.9. Two radical approaches to seat design: the forward tilting seat (left) and the kneeling chair (right).

design is that if you sit on the chair without thinking, you will tend to exert a backward thrust with the feet in order to stay in the seat—this is a particular problem if the chair is on castors. High-friction upholstery is not really an answer since female attire (in particular) generally provides a low-friction interface between the outer and inner garments—women, therefore, tend to slide out of their skirts. Experience suggests that balancing correctly on the forward slope seat is a skill which needs to be learned. According to Mandal (1981) users may take 1–2 weeks to get used to such chairs.

These difficulties should be overcome with the 'kneeling chair' which provides a seat sloping forward at some 30° to the horizontal, combined with a padded support for the knees. Brunswic (1981) evaluated these concepts. She found that:

(*i*) When seat angle and knee angle were independently varied, forward tilted seat positions did not result in a lumbar posture which was significantly different from that obtained with a horizontal seat and the knees at right angles.
(*ii*) The lumbar posture of subjects using a kneeling chair in writing and typing tasks was not significantly better than when they used a conventional office chair.

Drury and Francher (1985) evaluated a kneeling chair by means of a user trial of considerable sophistication. Subjects were typing or operating a computer terminal. They concluded that the comfort was "no better than conventional chairs and could be worse than well-designed office chairs". The principal complaints were difficulties of access-egress, pressure on the shins and discomfort in the knee region (due presumably to impairment of circulation resulting from the acute angle). There was "little or no decrease in back discomfort" and in spite of training subjects "often slumped forward to give a kyphotic spine".

The kneeling chair is rapidly increasing in popularity and I have not heard any complaints from its users (outside the context of laboratory trials). Its proponents (of whom there are many) argue that it is 'good for your back' because it helps maintain a lordosis. This does not seem to be the case in typing and writing—although it may be true in other activities (conversation, watching TV, etc.). Since most kneeling chairs lack a backrest this important source of support for the spine is lost altogether. Hence, the compressive force acting on the spine must be the same as it is in unsupported standing.

The evidence I have seen to date leads me to conclude that the kneeling chair does not offer any measurable advantage over well-designed and correctly adjusted chairs of the conventional type.

Chapter 9
Hidden dimensions—the perception of space

The physical characteristics of the spaces we inhabit also have psychological overtones. It is to these more elusive human factors, sometimes appropriately called 'hidden dimensions' (Hall 1969), that we shall now turn our attention.

Let us commence with a specific example. I am writing this chapter sitting at my desk in an upstairs room of a modest suburban house in North London. The height of this room, from floor to ceiling, is 2600 mm. Hence, when I stand up there is a clearance of some 970 mm above my head (ignoring suspended light fittings, etc.). For a 95th %ile male this clearance would be reduced to 800 mm. This might lead us to suggest that, even allowing for the occasional extremely tall person, at least 20% of the room's volume is wasted (except for some rather inaccessible storage space around the walls). There is little doubt, however, that this space above my head contributes in some way to my psychological well being—this may be an example of fallacy No.1 but I am fairly certain most people would agree with me. Now supposing, in the interests of space-saving, economy of heating, etc., an architect wishes to design living accommodation with lower ceilings. The current 'first preference' ceiling height for public sector housing in Britain is in fact 2350 mm (Tutt and Adler 1979)—in such a room a 95th %ile male will be able to just touch the ceiling with his fingertips when standing on tiptoes. Does the presence of an additional 250 mm of 'non-functional' headroom make a significant improvement in psychological well being? It is not easy to say but we may derive some tentative answers from the researches of Coblenz and Jeanpierre who studied the spatial impressions of subjects in an experimental room of variable dimensions (cited in Grandjean 1973) (Table 9.1). A slightly higher ceiling was preferred for both sizes of room when standing but the difference (60–70 mm) was small compared with the difference in height between a standing and sitting person (*c.* 400 mm), and no correlation could be detected between the size of the person and the size of the room he or she preferred. A slightly higher ceiling was preferred for both postures in the larger room but the difference was by no means great enough to maintain constant proportions between the height of the room and the lengths of the walls. A major imponderable of any research of this kind is the extent to which the experimental subjects can be considered 'reliable observers'. Are they, for example, influenced to a great extent by rooms with which they are familiar? We simply do not know. Some psychologists would argue that human

Table 9.1. Preferred ceiling heights of subjects in an experimental room.

Posture	Floor area (m^2)	Preferred ceiling height (mm)
Standing	12	2540
Sitting	12	2470
Standing	30	2700
Sitting	30	2640

Data from Coblentz and Jeanpierre as cited by Grandjean (1973).

experience is at best only an inference which can be drawn from observable behaviour and at worst a totally closed book. Most ergonomists maintain a cautious middle ground on this question and would seek to support the evidence of reported experience with observable behaviour and vice versa. (Sommer (1969) has discussed the matter at length.)

In considering the spatial experience and behaviour of human beings, two closely related key concepts emerge—'territoriality' and 'personal space'. The concept of territoriality was originally derived from observations of animal behaviour. Many species of bird and mammal will vigorously defend a home territory (the blackbirds and robins in my garden are pugnacious little brutes). By analogy, human beings (although they are an essentially gregarious species) may be described as showing signs of territoriality when they attempt to define a space as being for their own more or less exclusive use. In addition to private residences territories in this sense might include offices, areas surrounding a chosen seat in a public place (e.g., passenger conveyance, libraries) or even a favourite armchair. It is an unwritten law of social-spatial behaviour that territorial intrusion will be resented and should be avoided wherever possible. We learn at an early age that one simply does not sit down next to a stranger who is occupying a double bench seat on a bus or train unless all alternative seats are taken. Similarly, if we are the first occupant of such a seat we may choose to 'mark out' our territory by placing an item of luggage beside us to deter intruders.

Territory, then, can be temporary as well as permanent, which leads us to the concept of personal space—which has been compared with a 'psychological bubble' that surrounds us wherever we go and influences our interaction with other people. Sommer (1969) describes personal space as a portable territory—a region around the person's body, demarcated by invisible boundaries, into which the entry of other people is strictly controlled. Hall (1969) distinguished four concentric zones surrounding the individual. Each zone, defined in terms of face to face distances, was associated with a typical class of social interaction. The distances involved were believed to be determined by the characteristics of the sense organs and the lengths of the limbs as well as by cultural mores. In Table 9.2 Hall's original version has been modified by the substitution of round number metric equivalents and a paraphrasing of the more extensive descriptive material. Each

Table 9.2. The spatial zones of social interaction.

Zone	Phase	Distance (mm)	Features
Intimate	Close	<150	Total body contact—used for lovemaking, wrestling, and gestures of affection or comfort
	Far	150–450	Possibilities of body contact—reserved for intimate encounters
Personal	Close	450–750	A transitional spacing suggestive of intimacy, hence to stand at this distance at a social gathering is indicative of a close relationship
	Far	750–1200	Extends from an easy touching distance to a point where two people can touch fingertips, hence one can be 'kept at arm's length'. Used for communication of a personal kind
Social	Close	1200–2200	The characteristic separation of people at informal social gatherings or for groups of people working together. To look down on someone at this distance is authoritative or domineering
	Far	2200–3500	The distance to which people move when someone says "stand back so I can look at you". Used for business and social discourse of a formal nature. Offices may be laid out to keep people at this distance, e.g., by using a desk as a defensive feature. Similarly, intruders may be ignored without seeming excessively impolite and intermittent conversation may be interspersed with other activities
Public	Close	3500–7500	The voice must be raised and speech takes on an oratorical quality with careful choice of words and phrasing
	Far	>7500	At this distance one is making a public announcement, commonly of a theatrical kind, with exaggerated speech and gestures

After Hall (1969).

zone is divided into a close phase and a far phase which have slight differences of social meaning. We should not imagine that these zones have sharp transitions—naturally enough they merge into one another. None the less, they provide a model which can serve as a useful starting point for understanding spatial behaviour. The penetration of another person into a spatial zone inappropriate to the circumstances may be experienced as an unwanted intrusion of a stressful kind. The victim may respond to this stress in various ways, the most extreme of which is flight. Sommer (1969) documents examples of experimental studies of this phenomenon in environments ranging from geriatric institutions to university libraries. In circumstances where flight is inappropriate other responses may occur—hence, in the crowded rush-hour tube train where total strangers are thrust into the intimate zone of personal space, people stiffen themselves to minimize bodily contact and stare at the ceiling to avoid meeting the gaze of other passengers. In some cases the stress may be deliberately inflicted—hence the systematic invasion of personal space is a feature of certain police interrogation techniques.

Although the zone dimensions of Hall (1969) provide a starting point and model for the analysis of spatial behaviour, they should not be seen as fixed, immutable characteristics of human beings. Rather, they will vary both with the person and with the circumstances. Hall's figures were based upon observations of the social behaviour of Americans during the 1960s—a population which he considered to be middling in their social distances since, he noted, the hot-blooded Latin people characteristically approach each other more closely and the chilly British were more distant. However, it may well be that in the 20 years which have elapsed since Hall's observations, the breaking up of social taboos and hierarchies has led to an overall shortening of these distances; my impression is that the distances of Table 9.2 are large by present-day standards. There is also evidence that women have smaller psychological bubbles and, hence, tolerate closer encounters than men; that opposite sex pairs approach more closely than the same sex (especially if attracted); and that members of a peer group approach more closely than pairs of disparate age (Oborne and Heath 1979).

The psychological bubble is by no means spherical; we can better tolerate the approach of strangers side by side than face to face. Furthermore, we have vertical space requirements as well as horizontal. Savinar (1975) studied subjects in a small room with an adjustable ceiling, and found some evidence to suggest that the horizontal dimensions of personal space increased when ceiling height was reduced below that which was psychologically 'comfortable'. Similarly, Little (1965) found personal space to be dependent upon context; distances were greatest in an office waiting room and least outdoors.

Furniture may frequently act as an accessory or prop in the expression of territoriality, and as such may modify personal space and its associated behaviour. Figure 9.1 shows possible arrangements for furniture in an office. The left-hand figure could be characterized as 'defensive'—it is not calculated to put visitors at ease. The right-hand figure is much more welcoming to visitors and is more

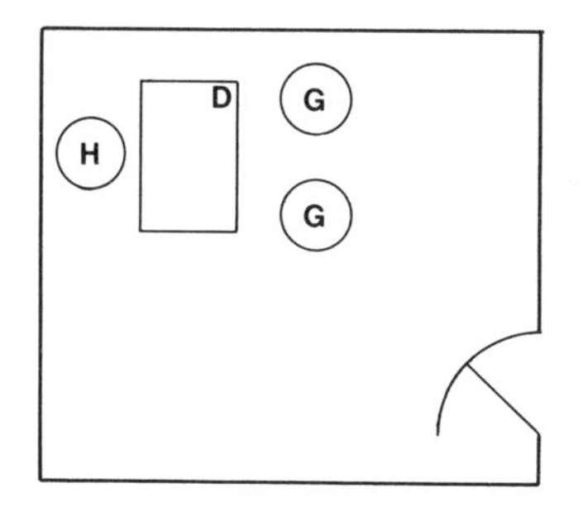

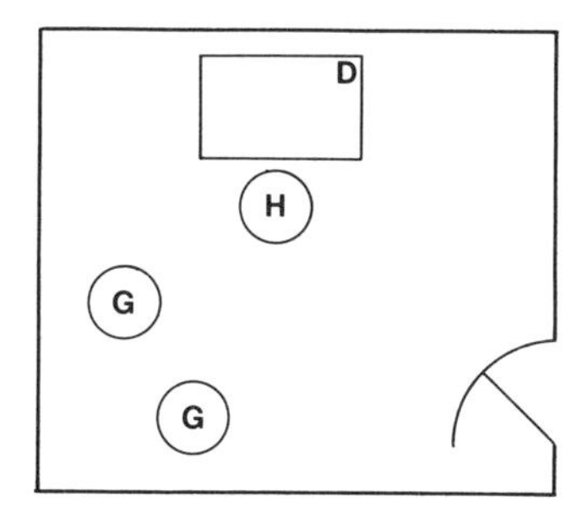

Figure 9.1. 'Defensive' (left) and 'relaxed' (right) arangements for furniture in an office. H = host seat; G = guest seat; D = desk. After Bennett (1977).

conducive to relaxed social interaction (Bennett 1977). In the defensive arrangement the desk is used as a 'barrier'—how high or wide must such a barrier be before it is of value as such? Although the question has never been studied experimentally one could hypothesize that a coffee table, typically between 350 and 450 mm in height, has minimal defensive value—or is the face-to-face distance created by the furniture the only relevant factor?

Many writers have discussed the concepts of 'sociofugal' and 'sociopetal' spaces (e.g., Sommer 1969). The former will discourage social interaction, the latter enhance it. In general, back-to-back seating arrangements or serried ranks of side-by-side seats will be sociofugal—airport transit lounges are the classic example. The critical feature is probably the likelihood of eye contact between occupants, since eye contacts commonly provide the cues for the initiation of conversation. The function of many public spaces is to be supportive of pre-existing social groups. These may be formally constituted as in the seminar room or board room, or informal as in restaurants or public houses. In either case it is important that the seating arrangements reflect both the nature of the likely interactions and the size of the groups. I sometimes frequent a particular public house in Islington which, being located beside the Regent's Canal, provides excellent opportunities for outdoor seating in good weather. The indoor area includes a relatively secluded side room with U-shaped seating arrangements, each of which is large enough for about eight people. Commonly these units become territories for couples or trios or occasionally quartets, since septets and octets are very rare social groups in such an environment. To enter a couple's territory in order to occupy the expanses of redundant seating is an invasion of privacy. James (1951) observed many thousands of people in public places and work situations—the results plotted in Figure 9.2 quite clearly show that small groupings predominate.

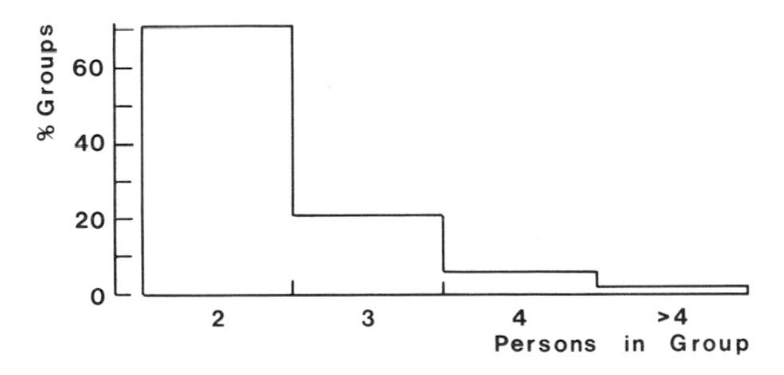

Figure 9.2. Frequency of occurrence of informal social groups of different sizes. Data from James (1951).

Chapter 10
People with disabilities

There is a very real sense in which every one of us is 'disabled'—since however we define or measure 'disability' we must do it in relative terms by comparison with some kind of average, norm or ideal state of health or functional competence. Logically speaking, a 'disability' is the absence of an 'ability' and if we chose to compare ourselves with Olympic athletes or musical virtuosi most of us would find that our 'abilities' were of an exceedingly modest scale. In general, we would expect to find, within the community at large, a continuum of any particular ability ranging from athletes or virtuosi at the one extreme to 'disabled people' at the other. It is a sad state of affairs that the English language seems to lack a non-pejorative word for describing the latter group of people. As Goldsmith (1976) observed, "any polite person knows that we do not call a person a cripple any more than we call him a bastard". To call someone an 'invalid', although seemingly less abrasive, is in a sense worse—since it carries with it the implication that the person's existence is in some way less valid than that of other people. A more neutral term like 'the disabled' is a violation both of logic and of the English language. Hence we are left with the rather cumbersome circumlocution 'people with disabilities'. This is not a euphemism—it is a plain description of the people concerned. Figure 10.1 summarizes the concept of the continuum of ability. The boundaries shown are fuzzy ones—we are all more or less athletic and all more or less disabled.

It is worth noting in passing that the World Health Organization defines health as a "state of complete physical, mental and social well being, not merely an absence of disease and infirmity". We can scarcely claim that this state is the norm or the condition of the average person.

It is perhaps informative to distinguish disability from its antecedents and consequences. Terminology is variable in this respect but an approximate consensus is given in Table 10.1 derived from Wood (1975), WHO (1980) and elsewhere. Hence the injury of a broken back will lead to the impairment of paraplegia (paralysis of the lower limbs), the disability of being unable to walk and the various social handicaps associated with life in a wheelchair (e.g., difficulty of access to buildings, public transport facilities, etc.). Those of us who are not sociologists may consider these semantic issues to be so fine as to approach triviality, but the distinction between disability and handicap is of the utmost practical

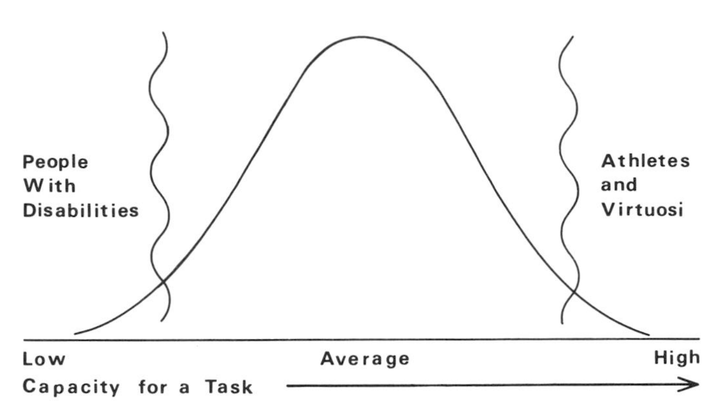

Figure 10.1. The continuum of ability.

Table 10.1. The semantics of disability.

Disease, injury, etc.
may lead to
Impairment—a disturbance of or interference with the normal anatomical, physiological or psychological conditions of the person
may lead to
Disability—the absence of certain abilities which the majority of people possess. A limitation of function or restriction of activity
may lead to
Handicap—social disadvantage consequent upon the previous stages

importance and is in many cases contingent upon the design of the environment. Streets without kerbstones, buildings without staircases and lavatories with adequate turning space do not alter the disability of the paraplegic but they reduce the burden of handicap imposed by the environment upon a person whose mobility is dependent upon a wheelchair. At the time of writing (December 1984) there is a very telling advertisement on the London Underground. It is issued by the Spastics Society and shows a man in a wheelchair at the top of a flight of stairs which clearly lead to a gentlemen's lavatory. The slogan reads "as far as I'm concerned it's neither public nor convenient".

10.1. *The incidence of disability and handicap*

The definitive survey of disability in Great Britain remains that of Harris (1971). The investigation was limited to people over 16 years of age, living in private households. A stratified sample of 2·5 million households were contacted by post and people who claimed some kind of impairment were interviewed. A checklist of activities necessary for independent living was used to measure the extent to which the person's impairment constituted a handicap. The definitions used in the survey differ somewhat from current usage, as given in Table 10.1, according to which the checklist would be deemed to measure disability. Extrapolating her findings to the population at large, Harris (1971) estimated that the number of adult people in Great Britain with an identifiable impairment was 3·07 million (or 7·8% of the adult population). Of these, 1·33 million (2·9% of the population) were handicapped—according to the checklist criterion.

Table 10.2 summarizes the data of Harris (1971) in the form of the age-specific incidence of disability. It comes as no surprise to discover that the incidence of both impairment and disability-handicap climbs steeply with age. Since we are a demographically ageing population we may expect the incidence of impairment and disability to steadily increase in the foreseeable future. Goldsmith (1976) has published some calculations of this kind. Using current census predictions (Ramaprakash 1984) we may estimate that the incidence of impairment will have climbed to 3·9 million (or 8·8%) by the turn of the century.

Table 10.2. Age-specific incidence of disability[a] (persons per 10 000 members of the age group concerned).

Age	Very severe disability	Severe disability	Appreciable disability	All disability levels	Impairment (with or without disability)
16–29	5	4	10	19	89
40–49	9	23	42	74	279
50–64	27	96	165	288	850
65–74	88	245	512	845	2207
75+	355	544	763	1662	3780
All ages	41	90	156	287	780

[a]The term disability is used here following the definition of Table 10.1—the original source uses the word 'handicap'.
Calculated from the data of Harris (1971).

An estimate of this kind assumes that the *status quo* will be maintained as far as the age-specific incidence of disability is concerned. The optimist might argue that the age-specific incidence of disability will decline as medical science advances and diseases such as multiple sclerosis or rheumatoid arthritis are 'defeated' or that changes in transport technology will reduce the number of road traffic accidents which lead to disabling conditions. The pessimist would argue that the triumphs of medical science up till now have been, largely speaking, in terms of keeping people alive—in many cases in spite of impairments and disabilities. If this continues to be the case, and the life expectancy of people with severe disabilities continues to increase, then our predictions may well be exceeded.

The survey of Harris (1971) dealt with people whose impairments and disabilities were of a more or less permanent sort. The incidence of the five most common categories of these impairments are shown in Table 10.3 (The original source does, however, point out that, due to the nature of the sampling methods, sensory or mental impairments are underestimated.)

Table 10.3. Principal categories of impairment—estimated numbers in Great Britain in 1969.

Category	Number	Percentage of all impairments
1. Disorders of the musculoskeletal system—osteoarthrosis, rheumatoid arthritis, backpain, fractures, muscular dystrophy, etc.	1 187 000	39
2. Disorders of the circulatory system—heart disease, arteriosclerosis, hypertension, etc.	492 000	16
3. Disorders of the central nervous system—cerebral haemorrhage (stoke), paraplegia, multiple sclerosis, poliomyelitis, cerebral palsy, Parkinsonism, epilepsy, etc.	360 000	12
4. Disorders of the respiratory system—bronchitis, emphysema, asthma, etc.	284 000	9
5. Disorders of the sense organs—blindness, deafness	277 000	9

After Harris (1971).

In general, these figures understate the problem of disability, since whenever we are laid low by disease or injury, to the extent that we cannot continue our accustomed range of activities, we are temporarily disabled. Low-back pain (see Chapter 8) is a case in point. Hult (1954) conducted a survey of some Swedish men working in a variety of occupations. He found that by the age of 60 years some 70% had been incapacitated by back pain (to the extent of taking time off work) for at least 1 day.

10.2. Design and disability

Data of the kind quoted above are of strictly limited value from an ergonomic standpoint, since a diagnostic term such as 'rheumatoid arthritis' only gives us a very vague impression of the patient's capacities and limitations. A means of cross-referencing between ergonomic and medical criteria is required if we are to extrapolate from a diagnosis to a design specification. With a few exceptions data concerning these matters are lacking; such information as exists is largely anecdotal based upon the collective experience of physicians and therapists.

People with the more extreme types of disability, such as the tetraplegic (with all four limbs paralysed) or the victim of severe brain damage, commonly require equipment customised to their own special needs. The *possum* environmental control system enables the paralysed person to control a range of domestic appliances (lights, television, door, intercom, typewriter, etc.) by means of whatever residual motor function he retains. This may involve a panel of micro-switches, a tube on which to suck and blow or even a microphone in which to whistle a control tone. Microcomputer technology (particularly involving voice input) has the potential to revolutionize the design of such aids.

In other circumstances 'low-tech' design may be equally valuable. For the brain-damaged child (or adult) a tolerably comfortable bodily support device may be the *sine qua non* for the development of even the most rudimentary communication or social skills.

These extreme cases, however, are the tip of a very large iceberg. As we consider lower and lower levels of disability, we are concerned with greater numbers of people whose capacities and limitations increasingly resemble the norm. The tetraplegic is close to the apex of a pyramid at the base of which are innumerable arthritics and back-pain sufferers. At the middle stages of this pyramid much can be achieved by relatively minor environmental modifications. The particular problems associated with wheelchair use will be discussed in the next section. Griew (1969) and Elmfeldt *et al.* (1981) cite numerous practical examples of ways in which an adaptation of workstation layout or equipment design allows the person with relatively severe disabilities to remain within the desirable socio-economic situation of productive employment. (The fact that this situation is at present inaccessible to millions of able-bodied people raises numerous questions outside the scope of this book.)

When it comes to the broad base of the disability pyramid the problems are of a different nature. Here we are dealing with 'almost fit' people who wish to undertake a full range of normal activities. They simply experience rather more pain and inconvenience in doing so than most other people would under the same circumstances. The arthritic or back-pain sufferer can act as a sensitive indicator of sub-optimal designs which go more or less unnoticed by the average user. Screwtop jars (and many other packaging media or domestic utensils) present insurmountable difficulties to the rheumatoid arthritic whose fingers are weak, stiff, deformed and painful—but in many cases the objects concerned are less than ideal for the rest of us. Shotton (1984) documented the problems experienced by rheumatism sufferers in the use of standard car seat belts. There is little doubt that if the deficiencies she identified were remedied, the use of seat belts would become easier for everybody—not just a minority group who may be conveniently labelled 'disabled'.

A similar argument could be advanced concerning back pain and many of the design criteria proposed elsewhere in this book (see Chapter 8 for example); sinks, worktops and handbasins which are too low, seats which fail to provide adequate postural support (especially in motor vehicles) are obvious examples. In these cases it is not that the rheumatoid or back-pain sufferer needs an object that is differently designed—it is rather that the rest of us have reserves of physical adaptability which enable us to tolerate design deficiencies. Looked at in this way many environmental adaptations or 'aids for the disabled' are only required as a consequence of faulty ergonomics. Why is it that, unlike almost any other kind of consumer artefact, 'aids for the disabled' are usually a dull grey-green in colour? Have you ever seen an orange wheelchair?

10.3. *Wheelchair users*

Goldsmith (1976) reported the results of a survey of wheelchair users in Norwich conducted between 1964 and 1968. Table 10.4 shows an extract of his results in the form of age-specific incidences of wheelchair use. Comparing them with the age-specific incidences of disability given in Table 10.2 we find that although wheelchair use increases with age, it does not do so as rapidly as disability in general. In this survey 65% of users were women.

Table 10.4. The age-specific incidence of wheelchair use.

	Wheelchair users	
Age	Number per 10 000 persons in age group concerned	Percentage of all people with disabilities in age group concerned
20–29	10	50
30–49	15	21
50–74	31	11
65–74	56	6
75+	132	8
All ages	33	11

Based on the data of Goldsmith (1976).

As Goldsmith (1976) has observed, the wheelchair user is handicapped three times over. Firstly, whatever condition put him in the wheelchair, the disabilities concerned will be handicapping in themselves. Secondly, he must operate at an eye-level which is some 400 mm lower than that of standing people, which is disadvantageous both physically and psychologically. Thirdly, he rolls around in a cumbersome, awkward, space consuming, distinctive and inelegant vehicle. Whilst the first handicap is outside the designer's control, the second and third relate to the traditional ergonomic considerations of reach, working level, clearance and access—as such they are potentially soluble. Ambulant people will find fixtures such as light switches, electrical outlets and window catches operable at any height from floor level to somewhere between 1800 and 2000 mm (although increasing difficulty will be experienced at the extremes). A wheelchair user (whose upper limbs are unimpaired) can reach a zone from around 600 to 1500 mm in a sideways approach but considerably less 'head on'. It may well be that the location of fittings within this limited zone will prove entirely acceptable for the ambulant users of the building, but in the case of working surface heights no such easy compromise is possible. Karagelis (1982), using a number of performance criteria, demonstrated that the optimal kitchen worktop height for wheelchair users was around 700 mm. Hamilton (1983), on the basis of a fitting trial, recommended a similar figure for the bottom shelf of an oven. Kitchen equipment installed at the 700 mm level for the benefit of a wheelchair user would be highly unsatisfactory for an ambulant spouse or helper. This problem is particularly acute in the case of elderly couples where one partner is a wheelchair user and the other, although ambulant, has diminished adaptive capacity (G. Stanton 1984, personal communication). The stationary wheelchair user occupies up to 1445 mm × 645 mm of floor space as against 380 mm × 630 mm for a standing person. He has a turning circle of between 1500 and 1700 mm diameter and requires a minimum clear width of 800 mm for motion in a straight line, whereas the ambulant person is generously accommodated at 650 mm and can squeeze through 400 mm without much difficulty (see Section 7.1). Swing doors can be a problem especially if heavily spring loaded and/or mounted in consecutive pairs with inadequate space between. Revolving doors are insurmountable obstacles as, of course, are steps and kerbs. An extensive collection of recommendations related to these matters is to be found in Goldsmith (1976). There is also a British standard, *Code c f Practice for Access for the Disabled to Buildings* (BS 5619). In an evaluation of the latter, Haigh (1984) concluded that only three out of its 17 major recommendations were based on identifiable empirical evidence.

It is not a simple matter to provide relevant anthropometric data for wheelchair users. Ideally, one should not only consider the variation of the user but also variations of the chair—a sampling strategy should be devised which is representative of both. If this were possible there would still be considerable difficulty. How, for example, should we interpret a percentile distribution of reach when a proportion of tetraplegic users cannot reach at all? The most informative survey of wheelchair users to date is the widely quoted study of Floyd *et al.* (1966) who measured 91 male and 36 female paraplegics and tetraplegics, the majority of whom were under 45 years old. In preparing Table 10.5 and Figure 10.2 the limited sources of data which are available have been ignored and the author has started from scratch, using the techniques of estimation given in Section 2.7. Estimates have been made on the following assumptions:

(*i*) Wheelchair users are a random sample of the population for any particular age group. Hence, their stature, were they able to stand, would be distributed accordingly.

(*ii*) The bodily proportions of wheelchair users (i.e., e_1 and e_2 as defined in equations 2.45 and 2.46, for estimation purposes) resemble those of the over 65 age group.

(*iii*) People sit in a wheelchair with a uniform slump of 40 mm.

Table 10.5. Anthropometric estimates for wheelchair users (all dimensions in millimetres).

Dimension	Men 5th %ile	Men 50th %ile	Men 95th %ile	Men SD	Women 5th %ile	Women 50th %ile	Women 95th %ile	Women SD
1. Floor to vertex	1260	1335	1410	45	1180	1265	1355	53
2. Floor to eye	1150	1220	1290	43	1080	1160	1235	50
3. Floor to shoulder	965	1080	1100	40	910	985	1065	47
4. Floor to elbow	625	685	745	37	610	670	730	36
5. Floor to knuckle	370	435	500	41	330	405	480	45
6. Floor to top of thigh	620	650	680	18	565	600	635	21
7. Floor to top of foot	120	150	180	19	165	190	215	16
8. Floor to vertical grip reach	1550	1665	1785	71	1460	1570	1680	67
9. Knee from front of chair	80	140	200	37	55	120	180	37
10. Toes from front of chair	360	435	505	43	305	370	435	40
11. Forward grip reach from abdomen	370	455	540	51	330	410	490	49
12. Forward grip reach from front of chair	250	315	385	41	175	240	305	39
13. Sideways grip reach from side of chair (shoulder–grip length)	580	645	710	38	520	580	640	37
14. Shoulder breadth (bideltoid)	390	445	500	33	330	380	430	30

Dimension	Minimum	Mean	Maximum
15. Overall length of wheelchair	915	1075	1445
16. Overall breadth of wheelchair	560	615	645
17. Height of armrests	705	735	770

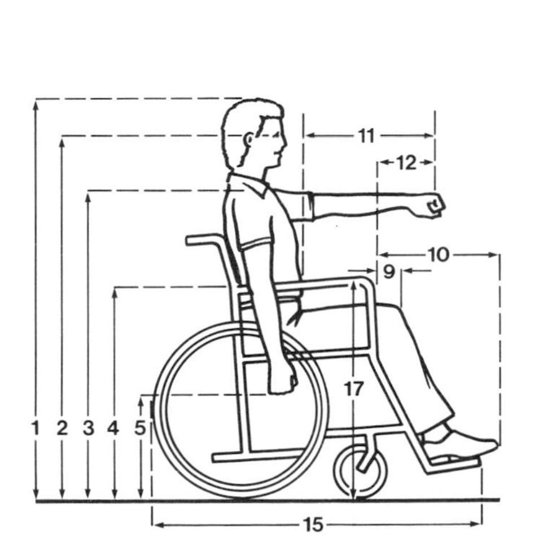

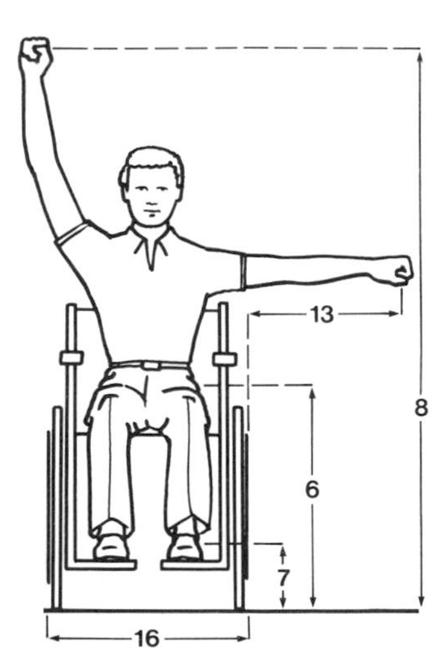

Figure 10.2. Anthropometric data for wheelchair users as given in Table 10.5.

(*iv*) The effective height of a wheelchair (including cushions) is 500 mm (G. Stanton 1984, personal communication).

(*v*) The horizontal distance from the SRP to the front edge of the chair is 440 mm (Goldsmith 1976).

Assumptions (*iv*) and (*v*) effectively assert that all dimensional variability is in the user and none in the chair. The reader should make his own corrections where this is demonstrably not the case. In calculating dimensions 7 and 10 it has been further assumed that the footrests of the wheelchair are positioned such that the knee is at an angle of 110° and that the user is wearing 25 mm heels. Dimensions 15–17 are quoted from data in Goldsmith (1976) for eight different makes of adult wheelchair.

10.4. Pregnancy

Pregnancy is a physiological condition which makes it difficult, if not impossible, to perform certain activities. It may, therefore, be considered as a disability within the limits of our definition. If we take the age-specific birth rate per annum and multiply it by 0·75 (since one is typically pregnant for nine calendar months) we obtain an approximate age-specific incidence of pregnancy (ignoring terminations, etc.). In the peak age group almost 10% of women are pregnant at any point in time (see Table 10.6). In the early stages of pregnancy the foetus is at risk from radiation and toxic hazards, and the lifting of heavy weights carries a risk of miscarriage. As pregnancy advances, bending and reaching activities are limited, standing for long periods becomes increasingly uncomfortable and physical activity in general becomes more arduous.

Table 10.6. Age-specific incidence of pregnancy (per 10 000 women in age group concerned; Great Britain in 1982).

Age	Pregnancy rate
15–19	210
20–24	765
25–29	945
30–34	518
35–39	165
40–44	38
15–44	449
All ages (over 15)	232

Data from Ramaprakash (1984).

I have only succeeded in locating one published anthropometric survey of pregnant women. Yamana *et al.* (1984) measured Japanese women from the second to tenth months of their pregnancy in order to obtain data for garment design. Table 10.7 is an estimate of two important anthropometric dimensions during pregnancy in British women (on the assumption that in proportional terms they change similarly to the Japanese).

Table 10.7. Estimated anthropometric changes of women during pregnancy.

	Month of pregnancy							
Dimension	2–3	4	5	6	7	8	9	10
1. Abdominal depth								
5%ile	195	210	225	250	275	300	315	330
50%ile	245	260	280	300	320	345	360	375
95%ile	295	310	335	350	360	385	405	425
SD	29	29	34	29	26	26	27	28
2. Forward grip reach from front of abdomen[a]								
5%ile	390	375	350	335	320	295	280	260
50%ile	460	445	425	405	385	360	345	330
95%ile	530	515	500	475	450	425	410	400
SD	42	42	46	42	40	40	41	42

[a]Note that dimension 2 is equivalent to dimension 36 minus dimension 21 in the general tables of Chapter 4.

10.5. The broad spectrum of users—a checklist

The concept of human diversity has been examined in detail elsewhere in this book. The present section concludes with a brief checklist of some of the people we might need to consider if we are to design for a broad spectrum of users—rather than the 90 or 95% who are clustered around the average.

(*i*) *Wheelchair users*. Has a suitable compromise been made between their needs and the needs of the ambulant majority?

(*ii*) *Children*. Is the product usable by children? If so, should it be?

(*iii*) *Handedness*. Has a suitable compromise been reached between the requirements of the 90% of people who are right-handed and the 10% who are left-handed?

(*iv*) *Anthropometric extremes*. Is the product-environment usable by:

(*a*) a woman in the advanced stages of pregnancy?

(*b*) an obese person, e.g., abdominal depth = 450 mm, hip breadth = 550 mm?

(*c*) an adult who is outside the arbitrarily defined clinically normal range of stature?

Consider the following hypothetical examples (Table 10.8):

Table 10.8. Adults outside the 'clinically normal' range of statures.

	Short	Tall
Stature	1200	2200
Elbow height	750	1380
Knee height	375	690
Popliteal height	305	555
Buttock–knee length	410	750
Buttock–popliteal length	340	625
Overhead reach	1420	2605

(Dimensions in millimetres obtained by taking extreme statures and average bodily proportions.)

(*v*) *Disability*. Is the product/environment usable by people with the following common problems:

(*a*) hands that are stiff and painful with a weakened grip?
hands that shake?

(*b*) inability to perform bending and twisting actions without intolerable discomfort or pain?

(*c*) weakness of the leg muscles, making it difficult to raise the weight of the body against gravity?

(*d*) impaired vision?

(*e*) impaired hearing?

(*f*) impaired mental function, loss of memory, confusion?

Chapter 11
Seating

11.1. Fundamentals of seating

The purpose of a seat is to provide stable bodily support in a posture which is:

(*i*) comfortable over a period of time;
(*ii*) physiologically satisfactory;
(*iii*) appropriate to the task or activity which is to be performed.

It is likely that a seat which is comfortable in the long term is also physiologically satisfactory and vice versa (see Chapter 8). The extent to which a seat achieves these objectives is dependent upon a number of anthropometric and biomechanical factors which we shall now discuss in a general way before considering particular cases (see Figure 11.1).

Seat height (*H*)

As the height of the seat increases, beyond the popliteal height of the user, pressure will be felt on the underside of the thighs. The resulting reduction of circulation to the lower extremities may lead to 'pins and needles', swollen feet and considerable discomfort. As the height decreases the user will (*a*) tend to flex the spine more (due to the need to achieve an acute angle between thigh and trunk), (*b*) experience greater problems in standing up and sitting down, due to the distance through which his centre of gravity must move, and (*c*) require greater legroom. In general, therefore, the optimal seat height for many purposes is close to the popliteal height, and where this cannot be achieved a seat which is too low is preferable to one which is too high. For many purposes, therefore, the 5th %ile female popliteal height (400 mm shod) represents the best compromise. If it is necessary to make a seat higher than this (e.g., to match a desk or because of limited legroom) the ill effects may be mitigated by shortening the seat and rounding off its front edge in order to minimize the under-thigh pressure. It is of over-riding importance that the height of a seat should be appropriate to that of its associated desk or table.

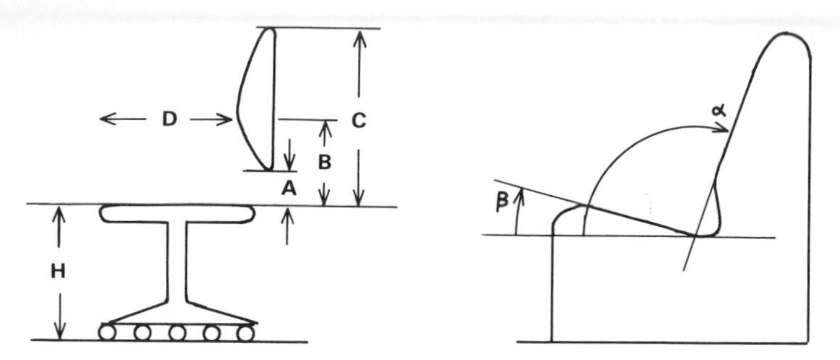

Figure 11.1. Seat dimensions given in the text and in Table 11.1.

Seat depth (*D*)

If the depth is increased beyond the buttock–popliteal length (5th %ile woman = 435 mm) the user will not be able to engage the backrest effectively without unacceptable pressure on the backs of the knees. Furthermore, the deeper the seat the greater the problems of standing up and sitting down. The lower limit of seat depth is less easy to define. As little as 300 mm will still support the ischial tuberosities and may well be satisfactory in some circumstances. Tall people sometimes complain that the seats of easy chairs are too short—an inadequate backrest may well be to blame (see below).

Backrest

The higher the backrest the more effective it will be in supporting the weight of the trunk. This is always desirable but in some circumstances other requirements such as the mobility of the shoulders may be more important. We may distinguish three varieties of backrest, each of which may be appropriate under certain circumstances.

Low-level backrest

For support to the lumbar region only. Ideally, this should commence at a level which clears the major protuberance of the buttocks, should have its maximum prominence in the mid-lumbar region and should conclude below the level of the shoulder blades to allow maximum freedom of movement for the shoulders and arms. Unfortunately the anthropometric data for these landmarks are inadequate—and the ergonomic and standards literature are a maze of conflicting recommendations and requirements. Rather than attempting to evaluate these in detail, Table 11.1 suggests specimen values by way of compromise.

The depth of the lumbar curve from front to back should be in the order of 15–20 mm.

Table 11.1. Specimen dimensions for a low-level backrest.

Dimension	Fixed (mm)	Adjustable (mm)
Seat to lower edge (*A*)	150	
Seat to maximum convexity (*B*)	230	170–300
Seat to upper edge (*C*)	380	
Height (*C–A*)		230
Width (for free movement of elbows)	330	330

Medium-level backrest

For full shoulder support a backrest height of 645 mm is required for the 95th %ile man. Such a backrest should have a forward convexity in the lumbar region (as above) which gently merges into a plane surface or concavity in the upper part. An excessively curved backrest is worse than one which is flat.

High-level backrest

Full support of the head and neck requires a backrest height of 900 mm for the 95th %ile man. For further discussion see Section 11.2.

Seat width

For purposes of support a width which is some 25 mm less on either side than the maximum breadth of the hips is all that is required—hence 350 mm will be adequate. However, clearance between armrests must be adequate for the largest user. The hip breadth of the 95th %ile woman unclothed is 435 mm. In practice, allowing for clothing and leeway, a minimum of 500 mm is required. (In some cases elbow–elbow breadth, as given in Table 5.3, is more relevant: 95th %ile clothed man = 550 mm.)

Backrest angle or 'rake' (α)

As the backrest angle increases, a greater proportion of the weight of the trunk is supported—hence the compressive force between the trunk and pelvis is diminished (and with it intradiscal pressure). Furthermore, increasing the angle between trunk and thighs improves lordosis. However, the horizontal component of the compressive force increases. This will tend to drive the buttocks forward out of the seat unless counteracted by (*a*) an adequate seat tilt, (*b*) high-friction upholstery or (*c*) muscular effort from the subject. Increased rake also leads to increased difficulty in the stand up/sit down action.

The interaction of these factors, together with a consideration of task demands, will determine the optimal rake which will commonly be between 100 and 110°. A pronounced rake (e.g., greater than 110°) is not compatible with a low- or medium-level backrest since the upper parts of the body become highly unstable.

Seat angle or 'tilt' (β)

A positive seat angle helps the user to maintain good contact with the backrest and helps to counteract any tendency to slide out of the seat. Excessive tilt reduces hip/trunk angle and ease of standing up and sitting down. For most purposes 5–10° is a suitable compromise.

Armrests

Armrests may give additional postural support and be an aid to standing up and sitting down. Armrests should support the fleshy part of the forearm, but unless very well padded they should not engage the bony parts of the elbow where the highly sensitive ulnar nerve is near the surface; a gap of perhaps 100 mm between the armrest and the seat back may, therefore, be desirable. If the chair is to be used with a table the armrest should not limit access, since the armrest should not, in these circumstances, extend more than 350 mm in front of the seat back. An elbow rest which is somewhat lower than sitting elbow height is probably preferable to one which is higher, if a relaxed posture is to be achieved. An elbow rest 200–250 mm above the seat surface is generally considered suitable.

Legroom

In a variety of sitting workstations the provision of adequate lateral, vertical and forward legroom is essential if the user is to adopt a satisfactory posture.

Lateral legroom

Lateral legroom (e.g., the 'kneehole' of a desk) must give clearance for the thighs and knees. In a relaxed position they are somewhat separated—a width of 600 mm is desirable but 500 mm would suffice (BS 5940 quotes a minimum of 580 mm).

Vertical legroom

Requirements will, in some circumstances, be determined by the knee height of a tall user (95th %ile shod man = 620 mm). Alternatively, thigh clearance above the highest seat position may be more relevant—adding the 95th %ile male popliteal height and thigh thickness gives a figure of 700 mm (BS 5940 quotes minima of 650 and 620 mm for general purpose and machine operators desks, respectively).

Forward legroom

This is rather more difficult to calculate. At knee level clearance is determined by buttock–knee length from the back of a fixed seat (95th %ile male = 645 mm). If the seat is movable we may suppose that the user's abdomen will be in contact with the table's edge. (Although, in practice, most people will choose to sit further back than this.) In this case clearance is determined by buttock–knee length minus abdominal depth, which will be around 425 mm for a male who is a 95th %ile in the former and a 5th %ile in the latter. At floor level an additional 150 mm clearance for the feet gives a figure of 795 mm from the seat back or 575 mm from the table's edge. All of these figures are based on the assumption of a 95th %ile male sitting on a seat which is adjusted to approximately his own popliteal height, with his lower legs vertical. If the seat height is in fact lower than this he will certainly wish to stretch his legs forward. A rigorous calculation of the 95th %ile clearance requirements in these circumstances would be complex but an approximate value may be derived as follows.

Consider a person of buttock–popliteal length B, popliteal height P and foot length F sitting on a seat of height H (as shown in Figure 11.2). He stretches out his legs so that his popliteal region is level with the seat surface (i.e., his thighs are approximately horizontal). The total horizontal distance between buttocks and toes (D) is approximated by

$$D = B + \sqrt{P^2 - H^2} + F \qquad (11.1)$$

(ignoring the effects of ankle flexion). Hence, in the extreme case of a male who is a 95th %ile in the above dimensions, sitting on a seat which is 400 mm in height

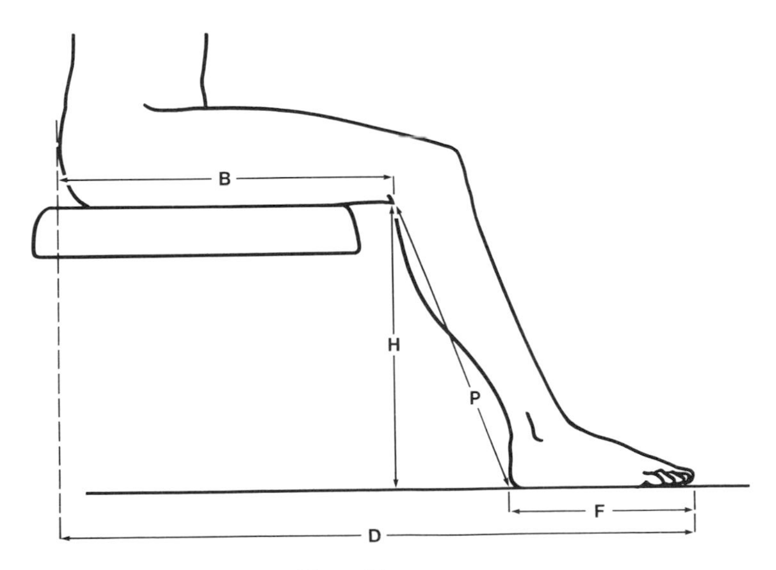

Figure 11.2. Calculation of forward legroom.

requires a total floor level clearance of around 1190 mm from the seat back or 970 mm from the table edge (if he is also a 5th %ile in abdominal depth). Such a figure is needlessly generous for most purposes; most ergonomics sources quote a minimum clearance value of between 600 and 700 mm from the table edge. (BS 5940 quotes minima of 450 mm at the underside of the desk top and 600 mm at floor level and for 150 mm above.)

Seat surface

The purpose of shaping or padding the seat surface is to provide an appropriate distribution of pressure beneath the buttocks. The consensus of ergonomic opinion suggests the following:

(*i*) The seat surface should be more or less plane rather than shaped, although a rounded front edge is highly desirable.
(*ii*) Upholstery should be 'firm' rather than 'soft'. (It is sometimes said that a heavy user should not deform it by more than 25 mm.)
(*iii*) Covering materials should be porous for ventilation and rough to aid stability.

The traditional wooden 'Windsor' chair can be surprisingly comfortable in spite of its total absence of upholstery. Its basic form was probably developed by the craftsmen of the Chiltern beechwoods sometime around the beginning of the eighteenth century. A critical feature seems to be the subtle contouring of the seat known as its 'bottoming'. This was hand carved, using first an adze then a variety of shapers, by a man known as the 'bottomer', whose specialised trade was considered the most skilled of all the activities which contributed to the chair-making process. He worked by eye without recourse to measurements —contemporary machine-made versions are said to be less satisfactory.

11.2. The easy chair and its relatives

The function of an easy chair is to support the body during periods of rest and relaxation. If not actually dozing or engaged in peaceful contemplation, the user may be reading, watching television or in conversation. The form of the chair follows naturally from these functions and from the considerations of the previous section.

Grandjean (1973) recommends a seat tilt (β) of 20–26° and an angle between seat and backrest of 105–110°. This gives a backrest rake (α) of as much as 136°, which is really only suitable for 'resting' and requires a degree of agility for standing up and sitting down. Le Carpentier (1969) found a tilt of 10° with a rake of 120° to be suitable for both reading and watching television. The present author inclines to the latter view with the caveat that for elderly users a rake of more than 110° may cause problems. Difficulties of standing up and sitting down will be reduced if the space beneath the front of the chair is unimpeded, allowing the user to place his feet beneath his centre of gravity, hence achieving a more vigorous upward thrust and a more controlled descent.

A high-level backrest (see p. 183) is virtually essential to the proper role of an easy chair in providing support for the trunk. Its shaping is something of a challenge. It is possible to design a gentle lumbar curve which will suit most users, but an equivalent pad for the neck and occiput is more problematic. Ideally, this should give you similar support to the natural action of clasping your hands behind your head. A sensible way of achieving this is to incline the upper part of the backrest forwards from the main rake by around 10° and to provide a movable cushion. (This solution has been adopted on certain British Rail seats but, unfortunately, the range of adjustment of the cushion is not quite adequate for the shorter person.)

The fundamental problems of designing an easy chair had essentially been solved by around 1680—as the collection of almost any English country house will testify. Ergonomic research has merely confirmed the intuitions of the designers of the past. However, the present-day furniture showroom typically presents a range of styles which, in ergonomic terms, are rarely better than just adequate and not infrequently fall short on numerous criteria. There are, of course, exceptions but these are commonly either reworkings of traditional types (such as the ever popular 'William and Mary') or else chairs which are described as 'orthopaedic' and sold more as 'aids' than as the furnishings of a stylish home.

The most common failings in the contemporary armchair are a seat which is too deep and a backrest which is too low. One may suppose that this is due to an attempt to make the seat and back equal in length in the interests of visual symmetry (like the Mies Van der Rohe 'Barcelona' chair of 1929) or to an even more misguided attempt to fit the entire chair into a cubic outline (like Frank Lloyd Wright's 'Cube' chair of 1895 or 'Le Grand Confort' by Le Corbusier and Charlotte Perriaud of 1928–1929). Combined with the weighty stylistic influence of these modern masters is a marketing need to incorporate the armchair into a three-piece suite (or some other combination). With the exception of a few historical types, such as the William and Mary 'love seat', high-backed settees are virtually unknown. In reality, as anthropometric data quite clearly show, the backrest height needs to be around twice the seat depth if an easy chair is to perform its proper function.

Tall people sometimes complain of seats being insufficiently deep (i.e., too short from front to back). Observation suggests that on engaging the backrest and finding that it only reaches mid-shoulder level, they move down into the seat in an attempt to gain head support. As a result their buttocks slide forwards until they are in danger of dropping off the front of the seat. (This also leads to the flexed position which is physiologically least satisfactory.) Hence the problem stems from an inadequate backrest rather than a seat which is not deep enough.

A common misconception, held by designers and consumers alike, is to equate depth and softness of upholstery with comfort. The luxurious sensation of sinking into a deep over-stuffed sofa is indicative of an absence of the support necessary for long-term comfort in the sitting position. In functional terms, we are now dealing with something more amorphous than a seat *per se*, it is in fact an object for sprawling or reclining on, rather than for conventionally sitting on. Structurally, however, the object retains the form of a seat. A seat supports its user in a sitting position and a bed supports him in recumbent position—but there are a whole variety of intermediate sprawling postures which can be perfectly satisfactory, especially when, supported by mounds of cushions, one has the opportunity for frequent postural changes. Taken to its logical conclusion the concept of 'amorphous furniture', which does not dictate any posture in particular, leads to items such as the 'sag bag'—a sack full of polystyrene beads, which enjoyed a brief vogue among young homemakers a few years ago. A whole family of all but extinct furniture types, which generically we could call couches, are essentially designed for sprawling—notable members of this family are the 'day bed' mentioned in Shakespeare (*Twelfth Night*, II.v) and the chaise-longue. A steeply raked easy chair can double as a couch when used in conjunction with a footstool—as in the ergonomically excellent Charles Eames lounge chair and ottoman of 1956 (Figure 11.3). The three-piece suite aims to serve for both sitting and sprawling. It commonly does both tolerably but excels at neither. There is considerable scope for design innovation in changing this state of affairs.

Figure 11.3. The Charles Eames lounge chair and ottoman (1956) give good support in a wide variety of postures.

11.3. Seats for more than one

When considering benches and other seats in which users sit in a row, it is necessary to bear in mind that the breadth of a 95th %ile couple is less than twice that of a 95th %ile individual. (The chance of two people, each 95th %ile or more, meeting at random on a bench is only 1 in 400.) In general, n people sitting in a row have a mean breadth of nm and a standard deviation of $s\sqrt{n}$, where m and s are the parameters of the relevant body breadth—which will usually be that of the shoulders (see p. 41). Table 11.2 gives values based on male data and including a clothing correction of 15 mm.

Table 11.2. Sitting in a row.

	Width required (mm)		
Number of persons	Mean	SD	95th %ile
1	480	28	526
2	960	40	1026
3	1440	48	1519
4	1920	56	2012

However, if the row of seats is divided by armrests the problem is more complex. Assume each user sits in the centre of his seating unit—a little reflection will tell us that the minimum separation of seat centres will be determined by the distribution of pairs of half-shoulder breadths: 480 [40] mm; 95th %ile = 545 mm. Since in the presence of armrests the minimum seat breadth is 500 mm (see p. 183), and an armrest cannot reasonably be less than 100 mm wide, 600 mm between seat centres will satisfy all criteria.

11.4. *The auditorium and its seating*

We shall conclude our discussion of seating by considering an example of seating accommodation within a particular space. In doing so we will, of course, go beyond questions of seat design *per se*. Seating arrangements in theatres and other auditoria pose an interesting collection of anthropometric problems. Not only is there the design of the individual seat unit (as discussed by Grandjean (1973)) but also the row problem as above and a rank problem *re* lines of sight, legroom and access. I know of one recently built London auditorium in which short people are uncomfortable because the seats are too high and tall people have inadequate legroom.

Let us commence by deciding that the height and depth of the seats will both be 400 mm (see pp. 181–182). Generous legroom for the 95th %ile man would be 1190 mm in front of the SRP (see p. 185) and the absolute minimum would be 645 mm, in which case his knees would touch the seat in front (providing his feet could fit beneath it). These figures result in row spacings (i.e., from one seat back or the SRP to the next) of 1390 and 845 mm, respectively, if the seat extends 200 mm behind the SRP.

Considering now the matter of access: two standing men, each of 95th %ile body depth, can pass crabwise in an unobstructed vertical of 760 mm (see Section 7.1). If a tilted seat takes 400 mm from front to back, a minimum row spacing of 1160 mm is required. The 'minimum' row spacing of 760 mm, cited in the 'AJ' *New Metric Handbook* (Tutt and Adler 1979), is extremely parsimonious. Trading off space against comfort, let us separate the rows by 1200 mm (600 mm for the seat, 600 mm for the gap).

What incline is required for unimpeded vision? This is partly an anthropometric problem and partly a geometric problem. If it were merely a matter of looking horizontally over the head of a person in front, the necessary increment would be determined by the difference between sitting height and sitting eye height (see Figure 11.4). We need to calculate a figure for a 95th %ile combination

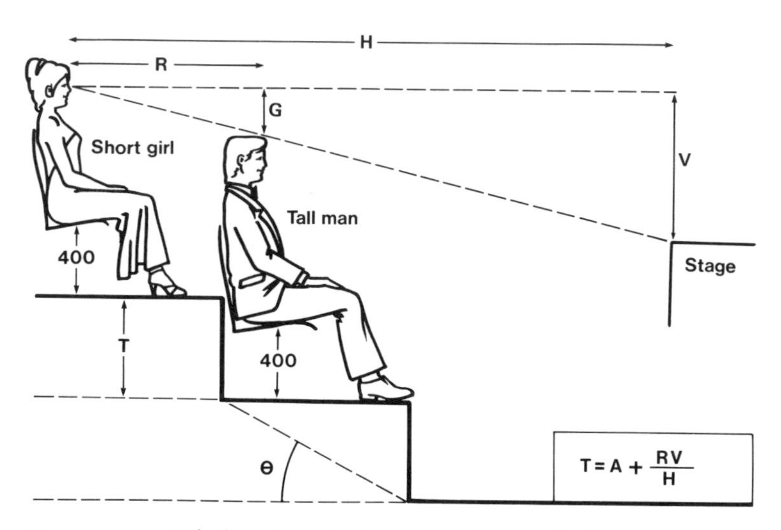

Figure 11.4. Sight lines in auditorium seating (see text for discussion), showing calculation of the geometrical increment (g).

of people. (See Equations 2.51 and 2.52.) If the sitting eye height of women is 740 [33] mm and the sitting eye height of men is 910 [36] mm, the distribution of differences is 170 [48] mm; 95th %ile = 250 mm. We shall call this figure the anthropometric increment (A). However, in most situations we do not have a horizontal line of sight to the stage but a downward one, and an additional geometric increment (G) must be taken into account. This latter is given by the equation $G/R + V/H$, where R is the between row spacing (as calculated above), V is the vertical distance of the viewer above the stage and H is his horizontal distance from it. Hence the total increment (T) between rows is given by:

$$T = G + A = RV/H + A \qquad (11.2)$$

Furthermore, the angle of rake (θ) of the floor is given by:

$$\tan \theta = T/R = V/H + A/R \qquad (11.3)$$

Fire regulations commonly require that the gradient of the aisle steps should not exceed 35° (Tutt and Adler 1979)—this is also the maximum preferred gradient of MIL-STD-1472C (Department of Defense 1981). To comply with this criterion and maintain unimpeded vision V/H may not exceed $\tan 35 - A/R$. If $A = 250$ mm and $R = 1200$ mm, then V/H may not exceed 0·49. Table 11.3 summarizes the results of these calculations for various conditions. If the rows of seats are staggered, then each person looks over another's shoulders rather than his head. The reader may wish to calculate the rake required if this is the case; the 95th %ile value of A is −70 mm in this case. Tutt and Adler (1979) give a graphical method for solving the gradient problem. It includes an anthropometric increment of 100 mm—being based on the fallacy of the 'average' man it will presumably result in impeded vision for rather more than 50% of people.

Table 11.3. Gradient (θ) *required for unimpeded vision in an auditorium as a function of row spacing* (R) *and location* (V/H).

	R (mm)				
V/H	800	1000	1200	1400	
0	17	14	11	10	Use ramps
0·1	22	16	17	16	
0·2	27	23	22	21	Acceptable for stairs
0·3	31	28	27	26	
0·4	35	32	31	30	Preferred gradient for stairs
0·5	39	36	35	34	
0·6	42	40	39	38	Acceptable for stairs in MIL-STD-1472C[a] but not in certain fire regulations
0·7	45	43	42	41	
0·8	48	46	45	44	
0·9	50	49	48	47	
1·0	53	51	50	50	
1·1	55	53	53	52	Definitely unacceptable

[a]MIL-STD-1472C considers gradients 20–50° to be acceptable for staircases but 30–35° is preferred.

Chapter 12
Office workstations

12.1. Desks and chairs

The basic configuration of office furniture is a chair and a desk at which the user will either write and perform other miscellaneous clerical tasks, or else operate the keyboard of a typewriter. It is generally accepted that for maximum comfort:

(*i*) the thighs should be approximately horizontal and the lower legs vertical, hence the seat should be at or a little below the popliteal height of the user;
(*ii*) for writing, the table should be at or a little above the user's elbow height;
(*iii*) for typing, the upper arms should be relaxed and hang vertically with the forearms horizontal, hence the home centre row of the keyboard should be at the user's elbow height.

The users are in contact with their environment in three places—floor, seat and desk/keyboard. It necessarily follows that the heights of two of these three should be adjustable if the exact prescribed posture is to be achieved by a range of users.

Virtually all contemporary office chairs are adjustable for height—although adjustable tables are beginning to make an appearance, in many ways it makes more sense to obtain the second degree of freedom by making the floor level adjustable, that is by provision of a footrest. At what height should the table be fixed? Grandjean (1981) concludes that for writing, etc., the optimal level is 270–300 mm above the seat (or 740–780 mm from the ground). This suggests that people prefer a writing surface around 50 mm above their elbows. BS 5940 specifies a height of 720 ± 10 mm (a figure which is more or less in line with most ergonomic references). Combining this recommendation with Grandjean's seat–table distance suggests that the optimum relationship will be achieved for seats adjusted between 420 and 450 mm. Since shod popliteal heights range from 400 mm (5th %ile woman) to 515 mm (95th %ile man) we might predict that the standard height is suitable for the shorter half of the population—or in practical terms that the taller user will be forced to adjust his chair to a somewhat lower level than he would choose in the absence of the desk. Cox (1984) confirmed that this was indeed the case. Subjects in a fitting trial identified their preferred combinations of seat and desk heights. The results suggested that desks ranging from 660 to 845 mm in height would be required to optimally match the full range of adult users. Should desks be higher in the interests of the taller person who has to crouch uncomfortably over the present standard height? The problem is that a desk which is high enough for a tall person makes a footrest virtually essential for the shorter user. At 720 mm, although ergonomic opinion might recommend a footrest, they are in practice virtually never used. The current standard is an expedient compromise—and in truth there is little evidence of dissatisfaction with it (Figure 12.1).

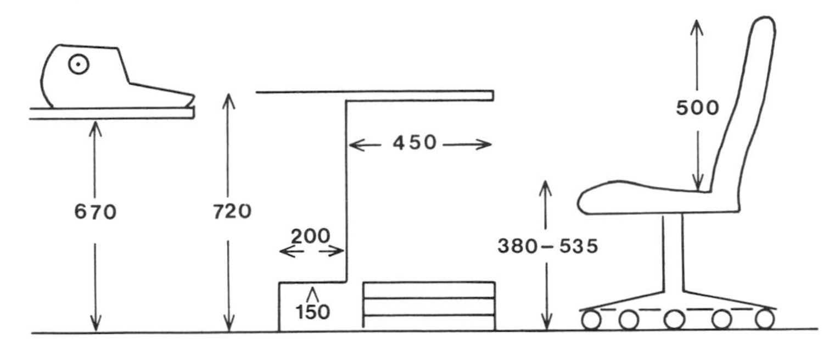

Figure 12.1. Compromise dimensions for office furniture (in mm)—see text for origin and derivation of figures.

Once we have settled on a standard desk height for writing, many other dimensions follow automatically. It is clearly sensible for the home row of a typewriter keyboard to be at the same height (so that the user can move from one to the other without re-adjusting the seat). BS 5940 recommends a height of 670 ± 10 mm for the typing desk; Grandjean recommends 650–680 mm.

Assuming that the correlation between sitting elbow height and thigh thickness is zero, there is a space of 80 [37] mm between the elbows and the tops of the thighs in women and 85 [34] mm in men. The home row of a typical typewriter stands some 70 mm above the desk top, which is itself some 25 mm in thickness. The inescapable conclusion is that the recommended typing posture is a physical impossibility for around half the population—irrespective of the design of the desk. The surprising narrowness of the gap between the elbows and thighs argues strongly in favour of low profile keyboards and strongly against drawers or other obstructions in the kneehole of the desk.

The range of adjustability of the seat is also determined in part by desk height. Given a 720 mm desk it is unlikely that anyone will want a seat higher than 535 mm (720 minus the 5th %ile female sitting elbow height). It is similarly unlikely that anyone will require a seat lower than the 5th %ile female popliteal height which comes to 380 mm in low heel shoes. (BS 5940 specifies an adjustment of at least 420–500 mm.) A woman who is 5th %ile in both respects will, in principle, require a 155 mm footrest—in practice she will probably settle for a compromise seat height.

Traditionally, typists' chairs have had low-level backrests and executive chairs have had medium- or high-level backrests. The supposed justification for this was that the typist required freedom of movement for the shoulders. (One does suspect, however, that the maintenance of 'status' and a puritanical distrust of comfort may also have been involved.) This convention has been questioned recently. Grandjean (1981) recommends a chair with a 500 mm backrest and positive lumbar pad (the apex of which he considers should be 100–200 mm above the lowest point of the seat). He cites trials in which such a chair was strongly preferred to a conventional typist's chair by users who had tried each for a period of 2 weeks. There was a slight preference for a version of the chair which tilted forwards by 2° and backwards by 14° (Figure 12.2). It may reasonably be argued that such a chair provides optimum back support in both the working and resting positions.

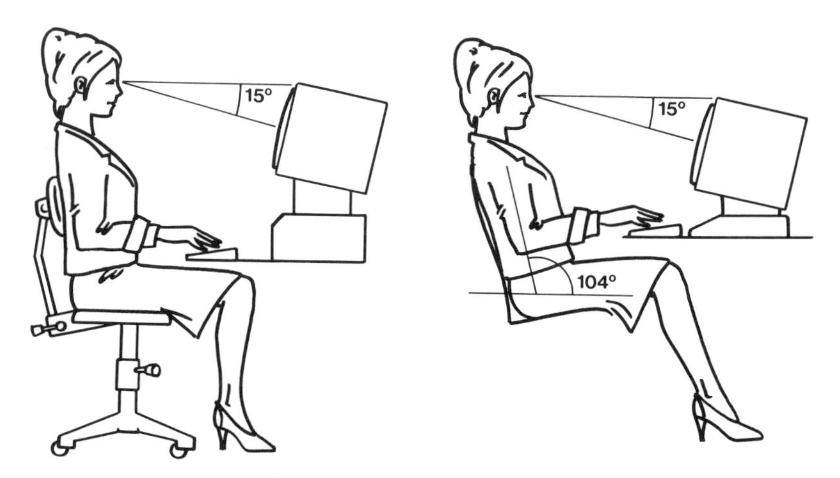

Figure 12.2. Working posture at the visual display terminal as recommended by Cakir et al. *(1980) (left) and Grandjean* et al. *(1982, 1984) (right).*

We have already noted the benefits to be derived from incorporating raised reading stands into the workstations of typists and VDT operators (see Section 8.4). Life and Pheasant (1984) have shown that they improve the work posture in the way one would anticipate and my experience is that, once having tried them, users find them indispensable. Similar arguments have been proposed in favour of a sloped writing surface. Once common, sloped desks have all but vanished from our schools and offices. Should the ergonomist mount a campaign for their re-introduction? Hira (1980) found that students using a tutorial room with adjustable desks preferred a slope of 10–15° to the horizontal. Bendix and Hagberg (1984) evaluated horizontal desks with slopes of 22 and 45°. They found that trunk posture improved with desk slope during a reading task (as we would anticipate). However, whilst the subjects clearly preferred the steepest desk (45°) for reading, they preferred the flat one for writing (although the latter effect was of marginal statistical significance). The best solution is probably either to incorporate the sloped writing surface into the desk as an optional feature under the user's control or else to make 'slopes' available as a separate item.

12.2. *The visual display terminal*

The visual display terminal (VDT) has in the last decade become, after the writing desk, the second most common category of workstation encountered in industry and commerce. There are those who believe that in the foreseeable future it will, with the advent of the 'paperless office', reach first place. The basic configuration which we will consider consists of a keyboard and screen. Other input-output devices will undoubtedly become widespread but to discuss them here would be premature. Important reference sources concerning VDT workstations include Cakir *et al.* (1980), Health and Safety Executive (1983), Ericsson (1983) and Grandjean and Vigliani (1980).

At present there are two schools of thought as to how the VDT workstation should be laid out (as shown in Figure 12.2):

(*a*) The majority view, as given in Cakir *et al.* (1980) and quoted in numerous guidance documents, advocates the same upright sitting posture as traditionally recommended for typing. Hence the keyboard should be at elbow height and the screen a little below eye level (as discussed in Section 8.4). Thus, standard office furniture, as described in the previous section, is considered to be satisfactory for VDT operators—provided that the keyboard and screen are appropriately designed (as described below).

(*b*) The minority view, which is rapidly gaining ground, has been proposed by Grandjean and his co-workers (Grandjean *et al.* 1982, 1984). In laboratory and field trials of an adjustable workstation it was found that, given a free choice, users will adopt a semi-reclining position with the keyboard substantially above the elbows. On average the trunk was at 104° to the horizontal, the upper arm at 113° and the forearm was inclined upwards by 14°. The latter is equivalent to a vertical distance at approximately 100–120 mm between elbow and fingertips. This posture has the effect of bringing the vertical levels of the keyboard and screen much closer together. A substantial majority of subjects were in favour of both keyboard and screen being independently adjustable for height, and the provision of a specially designed support for the wrists. (The latter presumably relieves the loading to the shoulder muscles which would otherwise result from this posture.) On the basis of these trials a keyboard height adjustable between 700 and 850 mm was recommended.

The following recommendations would, however, be generally acceptable (and are established beyond reasonable doubt):

(*i*) The optimum vertical position of the VDT screen is a little below the user's eye level (see also Section 8.4).

(*ii*) The keyboard and screen should be physically separate items allowing the user to locate them most advantageously for their own circumstances.

(*iii*) The screen should be adjustable in height with respect to the desktop and should preferably be tiltable so that the user can set it perpendicular to his own line of sight or to avoid glare from awkwardly placed light fittings.

(*iv*) A copy stand, at screen level, is desirable.

(*v*) The keyboard should be raked at 5–15° to the horizontal and made of visually unobtrusive, non-reflective materials. The vertical thickness of the keyboard should be minimized to reduce the clearance problems discussed in the previous section; hence the home row should not be more than 30 mm above the desk surface (see Section 8.6 for a further discussion of keyboard design).

(*vi*) The keytops should be 12–15 mm square with an 18–20 mm spacing between their centres. The keystroke should give auditory and tactile feedback with a depression force of 25–150 g and a travel of 1–5 mm.

If both seat and screen are adjustable the principal remaining issue is the optimal height of the desk on which the keyboard is placed. If this is also adjustable the choice becomes a problem for the user rather than the designer. At present, the market will perfectly well tolerate adjustable desks economically—but many ergonomists would argue that multiple adjustments in the workspace are less desirable than they seem, since they are commonly ignored or set incorrectly. The standard 720 mm desk does not actually prevent the user from adopting the posture chosen by Grandjean's subjects, but it would require a seat height of around 300 mm for the 95th %ile male or 440 mm for the 5th %ile female. Again we are forced to conclude that the standard desk is a compromise more suitable for the short than the tall. Finally, if VDT operators wish to recline somewhat, there is no doubt that a medium/high-level backrest is to be preferred.

Further discussion of keyboards and screens will be found in Chapter 17.

Office workstations—a personal viewpoint

In spite of extensive research into the design of office workstations (or perhaps because of it) a surprising number of controversies remain unresolved. Should the seat slope forwards as advocated by Mandal (1976, 1981) and discussed in Section 8.7? Should the writing surface slope? Should the desk be higher than the current 720 mm standard? Should it be adjustable? Should the VDT operator sit upright or recline? The evidence does not permit certainty but it would be unfair to the reader if I did not place my views on record.

Provisionally I am in favour of:

(*i*) incorporating a rocking mechanism into the seat which allows a few degrees of forward tilt as one option amongst many;

(*ii*) slightly raising the height of the standard desk to a level at or just above the midpoint of user preferences (760 mm is a reasonable figure);
(*iii*) making available a writing surface sloped at 15°;
(*iv*) the reclined position for VDT operation, especially with the use of wrist supports.

I am strongly in favour of raised reading stands and office chairs with a medium/high backrest (particularly for VDT operators).

I retain an open mind concerning adjustable desks and the radical proposals for the redesign of the keyboard described in Section 8.6.

Finally, there is little doubt that confining a person to a single desk or terminal throughout an 8-hour working day is undesirable. Changes of occupational activity and frequent rest pauses involving postural change and stretching exercises are all to be encouraged.

Postscript

There remains a persistent fear that VDTs might emit some form of radiation and that this might be potentially harmful to the unborn child. The general view, as typified by the Health and Safety Executive (1983), is that there is little or no cause for concern. None the less, recent figures from Japan (*New Scientist*, 27 May 1985) are disturbing. According to a survey conducted by the General Council of Trade Unions two-thirds of pregnant women working at terminals for more than 6 hours a day had problems with pregnancy or labour, whereas for 3–4 hours' work the figure was 46% and for less than 1 hour it was 25%. The causal link (if there is one) may be nothing whatever to do with radiation—but the problem cannot be ignored.

Repetitive strain injuries to the forearm are another cause for concern. Prolonged high-speed keying is generally blamed. It may well be, however, that the acceptability of a given workload will depend upon the posture of the wrist, which in turn is dependent upon the height of the keyboard with respect to the elbow—the higher the keyboard, the greater the extension and adduction of the wrist, the greater the strain. The findings of Duncan and Ferguson (1974) support this connection. It remains for further epidemiological studies to demonstrate whether workload or working posture is the more important factor.

Chapter 13
The driver's workstation

Worldwide, some 25 million new cars roll off the production lines each year. In Britain the motor industry contributes £12 billion annually to the Gross Domestic Product; 25 million people (approximately 60% of the eligible population) hold driving licences; there are some 20 million motor vehicles on the road and 84% of all passenger journeys are made by car (MIIS 1984). The resulting mayhem is considerable. On an average day 15 people die on Britain's roads (six occupants of cars, five pedestrians, three motorcyclists and one pedal cyclist)—an annual total of some 5500 deaths. The average British person has slightly better than even chances of being involved in a motor accident at some time; the chances of being seriously injured or killed are around 1 in 10 and 1 in 100, respectively. The *per capita* death rates are considerably worse in many of the European countries (Hamer 1985).

Can the application of ergonomics to the design of the driver's 'workstation' help reduce this problem—either in terms of primary safety (the prevention of accidents) or secondary safety (the reduction of injury)? Hard evidence is scant but at least one influential consumer is convinced. Attending an exhibition at the Design Centre in London, the Prime Minister, Mrs Margaret Thatcher, sat at the wheel of a prototype motor car and said "I don't like it, I cannot see the front. Redesign it for me . . . I like to see where the front of the car is so I don't bang into the back of a bus. If I was in insurance I would put up the premium" (*Daily Telegraph*, 27 March 1984). Mrs Thatcher is 1600 mm in stature—the 44th %ile for British women.

The layout of the driver's workstation, known in the industry as 'package design', poses complex ergonomic problems—both in terms of user diversity and the number of controls and displays involved. Furthermore, posture and seating must be such that a reasonable degree of comfort is maintained over extended periods. It is said that Alec Issigonis, designer of the Mini, deliberately made its seats uncomfortable in order to keep drivers alert! One cannot agree with this policy—it is more likely that an aching back will distract a driver than stimulate him.

Only a brief overview of the subject will be attempted here.

13.1. *Primary controls*

The driver's posture is largely determined by the location of the primary controls (i.e., pedals and steering wheel) in relation to the seat. What is a good posture for driving? This is very much a matter of compromise between conflicting requirements. It is desirable that the seat should be both raked and tilted (see Section 11.1) but an excessive trunk inclination will reduce the driver's visual field. A reasonable compromise might be a tilt of 10–15° and a rake of 105–115° (the latter is adjustable in most vehicles). In the interests of maintaining the lumbar lordosis (see Section 8.6) the hamstring muscles should be relaxed—requiring a relatively horizontal thigh and a knee which is flexed perhaps to a right angle. Such a posture would, however, be most disadvantageous for the application of force to the clutch and brake pedals (which are commonly operated with a thrusting action as against the pivoting action which is used for the accelerator). For maximal thrust the pedal should, in fact, be above the SRP at approximately the level of the hip joint and the knee should only be slightly flexed (Kroemer 1971, Pheasant and Harris 1982). In the latter study it was found that if the pedal is placed so that the user's instep is 12·5% of stature below the SRP and 47·5% in front, the available force is diminished by 12%. For a person of average bodily proportions this position is equivalent to a thigh angle 13° above the horizontal and a knee flexion of 66°. A pedal location which is somewhere near this point may, therefore, be deemed a suitable compromise (Pheasant and Harris 1982). The location of the steering wheel is probably less critical—it will be equally comfortable at close to arm's length or at half this distance. Furthermore, the wheel may be gripped in a variety of ways—in addition to the "10 to 2" recommendation of driving instructors. However, it is important that the wheel and column gives adequate clearance for the knees and abdomen. The former may be determined from the lower limb dimensions of a 95th %ile man with the seat in its rearmost position; the latter from the 95th %ile abdominal depth with the seat in its foremost position (or, even better, the abdominal depth of a pregnant woman).

How may these considerations be translated into actual dimensions? The classical approach, as described by Wisner and Rebiffe (1963) and Rebiffe *et al.* (1969), uses linkage anthropometry. It is possible to generate optimal zones for the location of hand and foot controls if we possess (*a*) dimensions for body linkages (as in Tables 6.1 and 6.2) and (*b*) postural criteria defining the optimal or acceptable limits for the angles subtended by adjacent linkages. Consideration of the manoeuvres required to bring the 5th and 95th %ile zones into alignment allows rational decision making concerning seat adjustability. An alternative but equivalent approach, which is arguably simpler, is to commence from the point where the driver's heel will rest (accelerator heel point, AHP) and work backwards to the location of his hip joint and, hence, to the SRP. Babbs (1979) has given a detailed description of a design method of this kind using mannikins.

Figure 13.1 shows a further analysis of these problems. Commencing from the AHP 5th %ile female and 95th %ile male lower limbs were drawn to scale in the

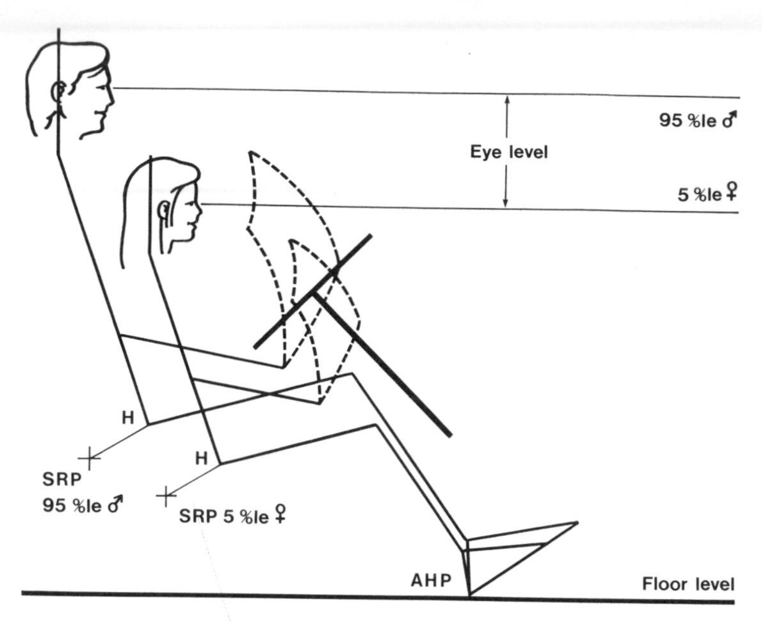

Figure 13.1. Layout of the driver's workstation using linkage anthropometry. Postural criteria as follows: thigh = 15°; knee = 70°; ankle = 0° (N); rake = 110°; shoulder < 45° Fl; 60° Fl < elbow < 110° Fl. (Thigh angle and seat rake measured to horizontal. Remainder according to the convention of Figure 7.7.)

same posture (knee angle 70°, thigh 15° above the horizontal) and the location of the SRP could be established for each. Hand grip zones were then added, such that the arm between the line of the trunk and an angle of 45° to the vertical and elbow angle is between 70 and 110°. Several interesting points emerge from the analysis.

If extreme members of the population are to adopt the same 'ideal' posture of the lower limb, the seat must have vertical adjustment as well as horizontal (or else adjust along an oblique track). However, if vertical adjustment is used in the manner shown, it will have the effect of increasing the disparity of eye levels. The eyes should ideally be around halfway up the windscreen—giving approximately equal views of road and sky. Alternatively, vertical adjustment may be used to equalise eye levels at the cost of deviating from the 'ideal lower limb posture'.

There is a degree of overlap between the hand zones of the 95th %ile man and the 5th %ile woman. A 360 mm steering wheel set at 45° has been drawn centred on this region. In this position it only just provides abdominal and knee clearance as described above. Hence, there is a strong argument for the adjustable steering columns which some manufacturers provide. Furthermore, a 360 mm diameter is at the lower limits of acceptability and 400–450 mm would probably be better, giving additional torque to the weaker driver. A more vertical wheel might be more easily accommodated. Lehman (1958), however, showed that the maximum force (torque) was exerted on a horizontal wheel, although paradoxically the speed of rotation was maximal when the wheel was vertical. He therefore recommended that the axis of the wheel (i.e., the steering column) should be 50–60° to the horizontal, in which position 70% of the maximum force can be exerted. At present there is a trend towards smaller and more vertical steering wheels, which are associated with a more 'sporty' image.

Decisions concerning control layout are, in practice, extensively constrained by factors unconnected with ergonomics. Hammond and Roe (1972) conducted a survey of a number of relevant dimensions and angles on more than 100 vehicle types, subjected the results to the statistical procedure of factor analysis and extracted a single 'package factor' (G) which identified some 85% of the variation between vehicle types. G increases with the height of the hip joint, wheel diameter, wheel angle and wheel height; it decreases with thigh angle, trunk inclination and the horizontal distance from wheel to heel. The distribution of G is not continuous; rather it clusters round modes for vehicle types. G is least for sports cars and increases through family saloons and light vans to heavy trucks.

Engineering constraints can have ergonomically unfortunate results. A prime example is the tendency, in small right-hand-drive cars, to offset the pedals and/or steering wheel slightly to the left of the seat—hence resulting in a twisted driving posture which is highly undesirable.

13.2. *Secondary controls*

The secondary controls range from critical, frequently used items, such as the gear shift, to non-critical, rarely used ones like the various climatic controls. Obviously enough they must be within the reach envelope of the smallest driver, with the most frequently used and most critical controls in the most accessible locations (e.g., on the steering column rather than the fascia). These envelopes are now the subject of an international standard (ISO 3958). The data are presented in the form of reach envelopes applicable to vehicles of a given range of package factors.

In addition to being within reach, the secondary controls should also be where the driver expects to find them. There is no particular location for the headlamp switch which is intrinsically superior—but the diversity of arrangements to be found in vehicles is a source of potentially hazardous confusion to which any driver could testify. The emergence of a standard for these matters (ISO 4040) is an important step in the right direction. Simmonds (1979) has described an interesting technique for evaluating the relative merits of different layouts. It is based on data for the frequency of use of controls (e.g., indicators) and the proportion of driving time that functions (e.g., headlamps) are in operation. Combining the two gives an index of likelihood of potentially hazardous errors (e.g., switching off headlights instead of operating indicators).

13.3. Standard aids to package design

The H-point, or average location of the hip joint in the compressed seat of a vehicle, is now officially defined by the use of standardized two- or three-dimensional mannikins. These were originally the subject of a Society of Automotive Engineers Standard (1974) and have subsequently been adopted in international standards (ISO 3958, ISO 6549). The H-point machine is a life-size three-dimensional model which is adjustable to represent a range of male body dimensions. Its back and thigh sections, when suitably weighted down, depress the car seat in a realistic fashion, allowing head clearance and lower limb angles to be measured directly. The H-point template is its two-dimensional equivalent.

Drivers' eye positions are the subject of a standard of considerable ingenuity (BS AU176, ISO 4513) which describes the construction and use of 'eyellipses'. These are elliptical contours in side elevation and plan which are not direct representations of the bivariate normal distribution of eye position (see pp. 25 and 37) although they are mathematically derived from it. Hence, the 95th %ile eyellipse does not include 95% of eye positions. Rather, it has the property that any line drawn tangent to its upper edge will have 95% of eye locations below it and 5% above it (and vice versa for the lower edge). In their standard form the eyellipses can only be employed in conjunction with the H-point devices —however, once this difficulty is overcome there is no reason why they should not have extensive application in fields other than automotive design.

13.4. The car seat

The angles and location of the car seat have already been discussed. Its dimensions may be derived from ordinary anthropometric considerations. It is generally agreed that the provision of good lumbar support is of great importance—it may well be that this can only really be achieved if the backrest contour is to some extent adjustable. Adjustability is even more desirable for the head/neck support—since if this is incorrectly located 'whiplash' may result in rear shunts. There is probably also some merit in seats which provide lateral support to stabilize the hips and back during cornering. Furthermore, the attenuation of vibration is an important consideration. The literature concerning car seats raises almost as many questions as it answers. None the less, there is little doubt that the car seats of today are, in general, very much better than those of a decade or so ago—but progress seems to be more by trial and error than otherwise.

13.5. Seat belts

If a seat belt is effectively to perform its function of restraining and protecting the driver (or passenger) in a crash, it must be wrapped around the strongest load-bearing structures of the body. Furthermore, it should be comfortable during general use. The diagonal strap should cross the rib cage, and apply its principal restraining force to the shoulder. It should neither slip off the shoulder nor engage or rub against the neck. The horizontal strap should wrap around the bony pelvis, without riding up over the abdomen where it might fatally injure the viscera. The fixing points of seat belts are (presumably) located to suit the average person. Anthropometric extremes commonly experience problems. Short buxom women are particularly affected since, because the alignment of the seat belt is too vertical, it rubs against the neck and compresses the left breast (in a right-hand seat).

The anthropometrics of seat-belt fit has been discussed by Haslegrave and Searle (1980) and Searle *et al.* (1980), who describe a sophisticated geometrical approach to the problem. The human body is represented by three geometrical forms representing the torso, thighs and legs. The seat belt follows a minimum distance path or geodesic across these forms. If surfaces of the forms are 'unwrapped' the geodesic can be drawn as a straight line. A computer model based on these concepts allows the evaluation of various seat-belt designs.

Chapter 14
Industrial workstations

The variety of workstations encountered in manufacturing industries is, of course, very great, and the greater part of this book may be applied in one way or another to their design. The present chapter deals with two particular topics concerned with safety.

14.1. *Safeguarding machinery*

The moving parts of machinery inevitably constitute a hazard. The traditional solution is the provision of guards which effectively separate the operator from hazard points where crushing, cutting, shearing, electrocution, etc., may occur. This poses a unique anthropometric problem since our usual criteria must be reversed—we wish to prevent access to hazard points (and/or place them out of reach). Tables 14.1 and 14.2 give relevant data which are derived from a German standard (DIN 31001, Pt. 1) via Clark and Corlett (1984). Consider the example of a machine which somewhere on its surface has a circular opening between 8 and 25 mm in diameter (e.g., for the lubrication of moving parts). Table 14.1 shows that such an opening will allow at least some operators to insert a finger—but no operators will be able to insert the whole hand. Any hazard point more than 120 mm away from this aperture will be harmless—since no operator's finger should reach that far. Table 14.2 shows that moving closer than the 120 mm safety distance will only trap and squeeze a finger if the gap which remains as they come together is less than 25 mm. BS 5304 contains an alternative set of data to Table 14.1 and BS 3042 describes a set of standard probes for checking hazards *in situ*. Davies *et al.* (1980a) on the basis of an anthropometric survey of women's hands concluded that the aperture dimensions given in both BS 5304 and DIN 31001 were "sufficiently small to prevent access to dangerous parts by all hands measured". The data of Table 5.5 confirm this. However, the 850 mm safety distance for an aperture which would admit the whole upper limb seems risky in the light of an estimated 95th %ile male limb length of 840 mm (Table 4.1); 900 mm would be safer.

When apertures exceed the largest sizes quoted in Tables 14.1 and 14.2, it is deemed that a certain proportion of users will be able to lean right in—hence the

Table 14.1. Safety distances when reaching through apertures.

Body part at risk	Aperture size[a] Square or circle	Slot	Safety distance to hazard points[b] (mm)
Fingertip	4–8	4–8	15
Finger	8–25	8–20	120
Hand (to root of thumb)	25–40	20–30	200
Whole upper limb	40–250	30–135	850[c]
Whole body	>250	>135	See Table 14.3

[a]Aperture dimensions given are the edge of a square, the circumference of a circle or the width (i.e., lesser edge) of a rectangular or elongated slot.
[b]Hazard points greater than the safety distance from the aperture are effectively isolated from the operator.
[c]This figure may be risky: 900 mm would be safer.
Data for DIN 31001 as quoted in Clark and Corlett (1984).

Table 14.2. Safety clearance for squeeze points.[a]

Body part	Safety clearance (mm)
Whole body	500
Lower limb	180
Foot[b]	120
Upper limb	120
Hand, wrist or fist	100
Finger	25

[a]A squeeze point does not constitute a hazard if a gap greater than the safety clearance remains between moving parts and if the next larger body part may not be introduced.
[b]As in object descending vertically onto foot.
Data from DIN 31001 as quoted in Clark and Corlett (1984).

safety distances will be much greater. DIN 31001 deals with this by tabulating the horizontal distance which a guard must be located from a hazard point as a function of the vertical heights of both (see Clark and Corlett 1984). Thompson and Booth (1982) have described an experimental evaluation of this standard. Male subjects of approximately 95th %ile stature were commonly able to reach further than the safety distances given in the standard. Table 14.3 is based, therefore, on the published data of Thompson and Booth (1982). The horizontal and vertical distances which the authors quote have been combined into a single radius of reach for each barrier height—it is, in fact, the greatest oblique distance reached for any of the hazard heights tested with that particular barrier. Since the true arcs of reach

Table 14.3. Safety distances when reaching over barriers.

Barrier height (mm)	Safety distance to hazard points (mm)
1000	1600
1200	1400
1400	1170
1600	940
1800	870
2000	590
2200	370

Data from Thompson and Booth (1982).

are not circular there will be certain directions in which the reach is less. It should be noted that the highest hazard position tested was 2400 mm above the ground. DIN 31001 considers a hazard located more than 2500 mm above the floor to be safely out of reach—again this is probably somewhat risky.

14.2. *Lifting and handling*

Analysis of the accident data published in the Annual Reports of HM Factory Inspectorate and (subsequently) the Health and Safety Executive reveals that, in every year between 1945 and 1980, 'handling goods' was cited as 'the primary cause' of between 25 and 31% of all reported industrial injuries. The equivalent pre-war figures, as far back as 1924 when records commenced, yield somewhat smaller percentages (HM Chief Inspector of Factories 1924–1974, Health and Safety Executive 1975–1980 a). Figures for other countries are similar.

There is a measure of uncertainty as to the distribution of body parts affected in these handling accidents. A breakdown for the year 1964 reveals a relatively even distribution: 37% to the trunk, 37% to the upper limb and 23% to the lower limb. Health and Safety Executive (1982) presents a different picture: 70% trunk, 19% upper limbs, 8% lower limbs (but the source of these data is not given). In the 1964 analysis 46% of accidents involving injury to the trunk had handling cited as their primary cause. Data for 1978–1980 onwards are classified in yet another way—'overexertion, strenuous or awkward movements and free bodily motion' account for 20·5% of all accidents in these years; 29·6% of accidents (of whatever cause) result in 'sprains and strains', and 15·9% of all accidents result in sprains and strains of the back (Health and Safety Executive 1978–1980 b). Troup and Edwards (1985) present a further analysis of these data: 61% of the overexertion injuries (12·5% of all accidents) involve the back and 74% of these back injuries were due to lifting actions. Hence, the particular category 'overexertion in lifting leading to back injury' accounts for around 9% of all industrial accidents. Finally, Hult (1954) (in Sweden) showed that around half of all episodes of back pain are associated neither with lifting actions nor with accidents but have a gradual onset. There remains a strong tendency to consider manual handling accidents and back injuries to be more or less co-extensive categories. This is not justified by the available evidence. Although the spine is at risk in handling activities it is by no means uniquely so.

In the half century covered by the figures there has been no reduction in the incidence of manual handling injuries—in spite of the overall 'lightening' of industrial work that is generally believed to have occurred. Research into the problem has been extensive and the number of words written on the subject must by now have passed six figures. A substantial proportion of this effort has revolved around two apparently simple questions:

Question A. What is the heaviest weight that is safe for a man (woman, child, etc.) to lift?

Question B. What is the safest method for lifting heavy weights?

Neither question is answerable—at best the answer of each is dependent upon that of the other and on numerous additional factors (such as the size and shape of the load, etc.).

The two questions that have commonly been asked typify the two classical approaches to accident prevention: the reduction of task demands by re-design or the enhancement of human capacity by selection and training. Which approach is more likely to prove effective? Snook *et al.* (1978), in an extensive survey of US industrial concerns, concluded that organizations which train their personnel in 'correct lifting techniques', or select them (on the basis of medical examinations, X-rays, etc.), experience no fewer back injuries than those who do neither. Were these just poor selection and training procedures? Perhaps so but one may search in vain for a documented account, of a controlled study, which unequivocally shows that a programme of selection and/or training has directly resulted in a reduced incidence of manual handling and/or back injuries. Snook *et al.* (1978) pointed to excessive task demands as the major identifiable causative factor in handling/back injuries and re-design of these tasks as the most promising solution.

In my own experience industrial manual handling tasks are characteristically 'undesigned' activities, and have something of an extempore quality. Human muscle power is commonly used to fill the gaps between sequences of machine operations in an otherwise automated process. The increased employment of industrial robots will not necessarily change this situation.

Workspace design

A very simple principle underlines most of our recommendations: that the centre of gravity of the load should at all times be as close as possible to the body of the lifter. This is justifiable on two separate grounds. Firstly, the closer the load the less its leverage about the articulations of the body; hence less muscular effort is required and the mechanical stress is less, particularly to the spine (as discussed in

Section 8.1). Secondly, the closer the load the more easily is it counterbalanced by the weight of the body; hence the load is less likely to get out of control. The interaction of these factors has been analysed in detail by Grieve and Pheasant (1982). Furthermore, if a burden such as a box or crate is hugged to the body, friction with the lifter's garments may help to support its weight. In practice, the closeness of the load is largely a matter of foot placement—specifically the elimination of obstacles which need to be reached over or reached into and which prevent the lifter from getting his feet beneath or around the load. Industrial pallets are a prime offender (particularly when circumstances require objects to be stacked in a far corner).

In addition to the above generally accepted principle, it is commonly held that lifting and turning actions, which impose a rotation or twist upon the spine, are particularly hazardous. In practice, these movements are often very difficult to eliminate, without doing away with the manual handling task altogether.

Given good foot placement subjects may be expected to exert their greatest lifting forces at around knuckle height (*c.* 700–800 mm). When the force must be exerted at a distance from the central axis of the body, either forwards, or sideways, the peak disappears (Figure 14.1). D. W. Grieve and A. M. Potts (1985,

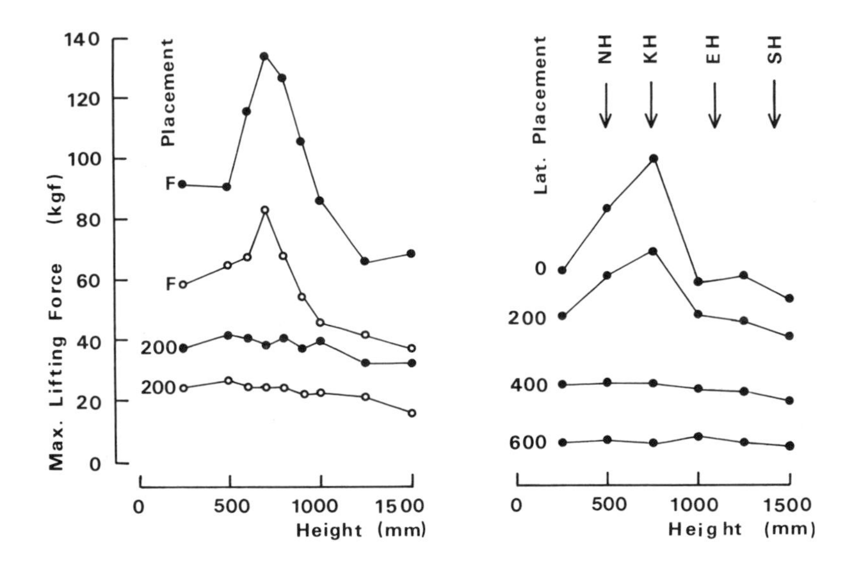

Figure 14.1. Strength of a static lifting action as a function of height above ground and foot placement. (left) Freestyle placement (F) and feet placed behind the axis of lift. Data kindly supplied by Anne-Marie Potts. ● = 16 men; ○ = 14 women.
(right) Feet placed 20 mm behind the axis lift 400 mm apart, various distances to the left. The placement figures are for the midline of the body. NH = knee height, KH = knuckle height, EH = elbow height, SH = shoulder height of the 21 male subjects. Data kindly supplied by Jane Dillon of the Furniture Industry Research Association.

personal communication) have shown that at the optimal range of heights 'cross-bracing' occurs, i.e., the elbows or forearms are supported against the thighs or trunk and the kinetic chain of the upper part of the body is stabilized. The knees and hip joints are slightly flexed and the lower limb muscles exert an extensor thrust, working into minimal leverage (Section 8.3).

What is the optimal height for lifting and handling tasks? To answer this question we must consider not only the initial exertion and the commencement of the lift but also its continuation to a successful conclusion. If the lift commences much below knuckle height the subject will either have to incline his trunk (hence imposing an additional load on the spine) or else strongly flex his knees (which reduces their mechanical advantage). A lift which is commenced at around knuckle height can, provided the load is not excessive, be continued comfortably to elbow height or a little more. If the load is a box or carton, which is gripped with both hands around its lower edges, difficulties will arise a short distance above elbow height since, as his wrist reaches its limit of adduction, the lifter will have to change his grip (or else make awkward compensatory movements of the shoulders and/or trunk). Lifts which commence at elbow height may be continued to shoulder height easily enough—but beyond this point the reduction in strength really begins to tell. When carrying a box or carton, it will again generally be held from beneath, and supported at hip height or above so as not to impede walking. Ayoub and McDaniel (1974) investigated horizontal pushing and pulling actions: on the basis of various criteria they concluded that the optimal height for pushing was 70–80% shoulder height and for pulling it was 70% shoulder height.

The following conclusions may therefore be drawn:

(*i*) Peak lifting forces are exerted at 750 mm above the ground.

(*ii*) For adult men and women, lifting box-like loads, heights between 650 and 1000 mm are most suitable.

(*iii*) Box-like loads will generally be carried between 800 and 1100 mm from the ground.

(*iv*) Lifts commencing at floor level should terminate no higher than 1100 mm but lifts which both commence and terminate in the height range 1000–1250 mm are also acceptable.

(*v*) The optimal height for pushing and pulling actions is around 1000 mm for men and 900 mm for women. Fixed horizontal handles for push-carts, trolleys, etc., should be at this level but vertical handles or movable horizontal ones allow the operator to find his own level.

The storage of items on shelving and racks is an important class of manual handling problem. In general, the heaviest or most commonly used items should be stored in the most accessible regions. Table 14.4 provides some guidelines based on the above criteria, anthropometric data for adult men and women and the experimental findings of Thompson and Booth (1982).

Table 14.4. Recommendations for the design of storage, shelving and racks.[a]

Height (mm)	Application and comments
<600	Reserve storage for rarely required items. Fair accessibility for light objects; poor for heavy
600–800	Fair for heavy items; good for light items
800–1100	OPTIMAL ZONE FOR STORAGE
1100–1400	Fair-good for light items—visibility unimpeded, access fair. Poor for heavy items
1400–1700	Limited visibility and accessibility. Most men and women will be able to stow and retrieve light items (at least on the edge of the shelves)
1700–2200	Very limited access; beyond useful reach for some people
>2200	Out of reach for everybody

[a]For present purposes, loads greater than 10 kg are deemed 'heavy'. Shelf depth should not exceed: 600 mm at heights of 800–1400 mm; 450 mm at heights less than 800 mm; 300 mm at heights greater than 1400 mm.

Minimum acceptable unobstructed space in front of the shelves: 680 mm for small items at heights greater than 600 mm; 1000 mm at heights less than 600 mm or for bulky items at any height.

Container design

Items for lifting and handling are commonly packaged into containers such as boxes, crates, trays, etc. Drury (1980) has reviewed the design of these. The following guidelines are based upon his conclusions but figures have been changed for compatibility with present anthropometric data:

(*i*) The container should be as compact as possible so that its centre of gravity is as close as possible to the subject's body.

(*ii*) Unpredictable or unexpected inertial characteristics should be avoided; hence the centre of gravity should be near to the geometrical centre of the container and the contents should be stably packed so that the centre of gravity does not shift around.

(*iii*) For loads which are to be lifted from the ground (which is undesirable) it is advantageous for the width to be less than 300 mm, allowing the object to pass between, rather than in front of, the knees.

(*iv*) Loads should not block vision when carried at hip height—hence their maximum height should not exceed the 5th %ile distance between hip and eye height (645 mm for men, 630 mm for women). If obstacles are to be negotiated at floor level, a downward sight line is required and load heights should be reduced accordingly.

(*v*) Furthermore, it should be possible to reach the bottom front corners of the load when carrying it at hip height. Its depth (front to back) should not exceed approximately 400 mm for men or 300 mm for women. (These figures are substantially less than the equivalent ones quoted by Drury (1980).)

(*vi*) If the load is to be carried at the sides (as for a suitcase) it should clear the ground when the carrier stands erect. The height of the case (including handle) should not, therefore, exceed knuckle height less 50 mm (5th %ile man = 665 mm, 5th %ile woman = 635 mm; allowing 25 mm for shoes in both sexes). Loads carried in this way should be as slim as possible.

(*vii*) Handles or hand-holds are a great advantage on crates, etc. These should be near the top of the crate for loads to be lifted from floor to table height. For loads to be handled at higher levels they should be lower down the crate.

(*viii*) Hand-holds should allow 110 mm to clear the breadth of the palm and 45 mm for the knuckles. A cylindrical handle approximately 40 mm in diameter is preferable (see Chapter 16).

Acceptable loads for lifting

The maximum load which men or women can reasonably be called upon to lift is dependent *inter alia* upon the factors we have discussed above. Guidelines will of necessity be complex but it should be possible to specify a load which is not acceptable for anybody under any circumstances, a load which is acceptable for anybody under any circumstances and some rules which will aid our decision making in the grey area between these extremes. The various methods which might be used to validate such recommendations have been reviewed elsewhere (Grieve and Pheasant 1982, Chaffin and Andersson 1984, NIOSH 1981, Troup and Edwards 1985). The present discussion will be limited to ends rather than means.

The currently favoured code of practice in the USA is that of NIOSH (1981), which sets a maximum permissible limit (MPL) and an action limit (AL) which are calculated as follows:

$$\text{AL (kg)} = 40(150/H)(1 - 0{\cdot}0004|V - 750|)(0{\cdot}7 + 75/D)(1 - F/F_{max}) \qquad (14.1)$$

$$\text{MPL} = 3\ \text{AL} \qquad (14.2)$$

where H is the horizontal location of the load, forward of the midpoint between the ankles at the origin of the lift (mm), V is the vertical level at which the lift commences (mm), D is the vertical distance that the load travels (mm), F is the average frequency of lifts/minute and F_{max} is the maximum frequency which may be sustained. H is deemed to range between 250 and 800 mm, V from 0 to 1700 mm and D from 250 to 800 mm. When $V + D/2$ is greater than 750 mm, F_{max} is 18 for 1 hour or 15 for 8 hours. When $V + D/2$ is less than 750 mm, F_{max} is 15 for 1 hour or 12 for 8 hours. For lifting less frequently than 1 per 5 minutes set $F = 0$.

The MPL is held to be equivalent to the capacity of a 75th %ile man or a 99th %ile woman; the AL to be equivalent to a 1st %ile man or 25th %ile woman. Workloads above the MPL are deemed unacceptable and should be reduced and those below the AL are deemed to represent 'nominal risk'. Between MPL and AL selection and/or training of the workforce is required. The values given by these equations are compatible with the strength data of Figure 14.1 and with the principles of workspace design outlined above. It is noteworthy that the vertical height factor in the equation is dependent upon the absolute distance from 'knuckle height' regardless of direction, i.e., $V - 750$. This correction is still applied when the value of H is such that the knuckle-height peak is not present. However, since the correction for H is generous this is probably of minor concern. Figure 14.2 shows action limits for occasional lifts in the optimal and extreme height range.

A major drawback of the NIOSH scheme is that it only covers symmetrical two-handed lifts, performed directly in front of the body—in practice, such tasks are relatively uncommon. (In my experience, and that of many colleagues with whom I have discussed the matter, most manual handling tasks on the shop floor involve sideways movement, rotation or some other asymmetry.)

Proposals for a code of practice published by the UK Health and Safety Executive (1982) are simpler but much less explicit. They specify thresholds at 16, 34 and 55 kg at which it becomes increasingly necessary to select and train handling personnel—or else provide mechanical lifting equipment. These threshold levels are held to represent the capacity of the 10th, 80th and 90th %ile healthy adult (in the last case with appropriate training) working "in idealized circumstances" (*sic*). Attention is drawn to those 'abatement factors', such as the age or sex of the workforce or the design of the workspace, which would lead one to modify the threshold levels.

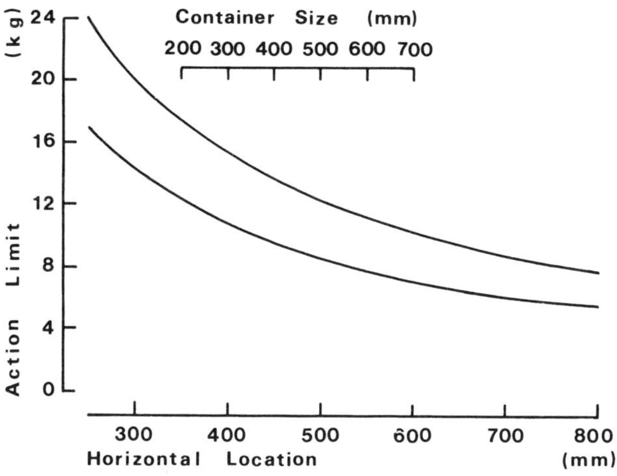

Figure 14.2. Action limits for occasional lifting actions calculted from Equation 14.1 plotted against the horizontal location of the load (H). *Lifts of 250 mm starting 750 mm from ground (upper curve) and either ground level or 1500 mm (lower curve). The front to back dimension of a container* (C) *equivalent to a given value of* H, *was calculated on the assumption that its nearside edge is held 250 mm in front of the ankles and that its centre of gravity is located at its geometrical centre. hence* H = C/2 + 250.

It is somewhat disturbing that both NIOSH (1981) and the Health and Safety Executive (1982) extensively advocate selection and training as approaches to the solution of manual handling problems when, according to the findings of Snook *et al.* (1978), such means are unlikely to be efficacious in reducing the incidence of injury. Furthermore, the highest possible value of MPL given by the NIOSH (1981) equations, representing the maximum permissible load for a trained 75th %ile man lifting under optimal conditions, is actually 120 kg—more than double the supposedly equivalent figure of 55 kg given in Health and Safety Executive (1982).

The most comprehensive set of recommendations to date are those of Davis and Stubbs (1977, 1978); recently adopted as a Ministry of Defence (1984) Standard (MoD 00–25, Pt. 3). These are based on an extensive programme of research which used the pressure within the abdominal cavity (intra-abdominal pressure, IAP) as an index of stress to the trunk and spine. A criterion IAP of 90 mmHg was set on the basis of epidemiological studies which showed a high incidence of back injuries in jobs in which this level was frequently exceeded. Contour maps in elevation and plan are given, showing the lifting, pushing and pulling forces which will not exceed the criterion in 95% of the population (Figure 14.3). A wide range of one- and two-handed exertions in standing, sitting and kneeling positions are covered in the original sources.

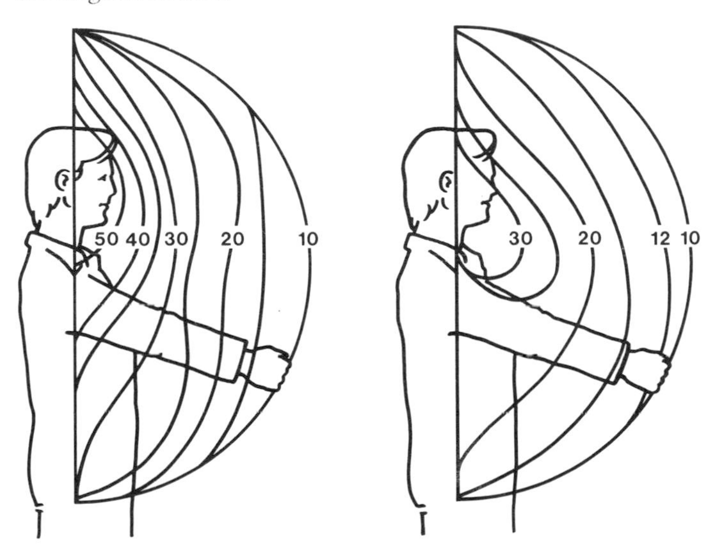

Figure 14.3. Suggested limits for lifting forces (kgf) two-handed (left) and one-handed (right) at various positions in the zone of convenient reach (ZCR) as given by Davis and Stubbs (1977, 1978) and Ministry of Defence (1984). Values given are for men under 50 lifting less than once per minute.

Table 14.5 is based upon the contour maps of Davis and Stubbs (1977, 1978) and MoD 00–25, Pt. 3. The loads may seem small by industrial standards. If so, that is because they represent the capacity of a nominal 5th %ile person. A majority of

Table 14.5. Maximum acceptable forces in manual handling activities[a] (all figures are given in kilogram force).

Activity[b]	Men				Women			
	Under 50		Over 50		Under 50		Over 50	
	Occ.[c]	Freq.[c]	Occ.	Freq.	Occ.	Freq.	Occ.	Freq.
Lift								
Two-handed; compact load, close to the body, within preferred ranged of heights	30	21	24	14	18	13	14	10
Two-handed; bulky load or compact load at half arms length or compact load outside preferred range of heights	20	*[d]	16	*	12	*	10	*
Two-handed at arms length or close to the body overhead	10	*	8	*	6	*	5	*
One-handed, compact load close to the body	20	14	12	8	12	8	7	5
One-handed at arms length either forwards or sideways	10	*	6	*	6	*	4	*
Push								
Two-handed above head	12	*	10	*	7	*	6	*
Two-handed shoulder height	25	18	20	14	15	11	12	9
Two-handed optimal height	45	31	36	25	27	19	22	15
Pull								
Two-handed above head	22	*	18	*	13	*	11	*
Two-handed shoulder height	35	25	28	19	21	15	17	12
Two-handed optimal height	50	35	40	28	30	21	24	17

[a]All tabulated data are taken from Davis and Stubbs (1977, 1978) and Ministry of Defence (1984), to which the reader is referred for further information. The loads given are deemed acceptably safe for 95% of the population of the age and sex specified. However, THE RESTRICTION OF LOADS TO WITHIN THESE (OR ANY OTHER) LEVELS DOES NOT GUARANTEE SAFETY.

Loads beyond the capacity of one person, may, if circumstances such as space permit, be handled by two. However, this does *not* double the acceptable load since the two persons' efforts will not be perfectly co-ordinated. For two operators, *increase* the load by between 50% and 80% depending upon the difficulty of the circumstances.

[b]It is assumed that the activity will be performed in a reasonably upright standing position. Where circumstances such as reduced headroom prevent this, loads should be reduced by 60%. Two-handed lifts are deemed to take place in front of the body; when this is not the case, loads should be reduced by around 20%. See Ridd (1985) for a further discussion of these matters.

[c]Occasional actions performed less often than once per minute, frequent actions more often.

[d]Actions marked with an asterisk are considered by the present author to be unacceptable. Thus no figure is quoted.

people will, therefore, be able to handle heavier loads—in some cases up to the somewhat arbitrary limit of 55 kg (Health and Safety Executive 1982). The unresolved problem is how to identify those individuals at risk in any given situation and direct them into lighter activities.

I should like to conclude by expressing a personal viewpoint: there is no such thing as a safe load. An unfit person may injure his back (or more accurately trigger an attack of pain in his already degenerate spine) by reaching awkwardly to pick up the most trivial of loads. A fit person may injure himself handling a very modest load if he slips and loses his footing. It is not possible to specify a load which guarantees safety. Such a viewpoint will, of course, be unpalatable to those involved in litigation.

Chapter 15
Domestic workstations

Every home has a space in which food is prepared, a space for eating, a space for sleeping, etc. Although these spaces may physically overlap it is more conventional, in Western society at least, for each room in a house to be the focus of a specialised range of purposive activities. In terms of numbers and volume of production these domestic workstations (and the equipment used within them) far outrank anything encountered in industry or commerce. Extensive discussions of the ergonomics of domestic work are found in Steidl and Bratton (1968) and Grandjean (1973). Design guides include Department of the Environment (1972) and Noble (1982).

15.1. *The kitchen*

The literature on this subject reveals two basic design principles which turn out to be based on sequence of use and frequency of use, respectively (see Section 17.4).

(*i*) For a right-handed person the sequence of activity proceeds from left to right thus: sink to main worksurface to cooker (or hob) to accessory worksurface for 'putting things down'. It clearly makes sense for this sequence to be unbroken by tall cupboards, doors or passageways; but it need not be in a straight line—an L- or U-shaped configuration will serve just as well. Another accessory worksurface left of the sink completes the layout (Figure 15.1).

(*ii*) The refrigerator (or other food store such as larder, freezer, etc.), sink and cooker constitute the much discussed 'work triangle' of frequently used elements. For reasons of safety, through circulation should not intersect this triangle—particularly the route from sink to cooker which is used more than any other in the kitchen. The sum of the lengths of the sides of the triangle (drawn between the centre front of the appliances) should fall within certain prescribed limits. Grandjean (1973) quotes a maxima of 7000 mm for "small to medium sized kitchens" or 8000 mm for "large kitchens". Department of the Environment (1972) gives a minimum combined length of 3600 mm and a maximum of 6600 mm for the kitchen "to leave adequate working space and yet be reasonably compact" and also specifies that sink/cooker distance should be between 1200 and 1800 mm in length (Figure 15.1).

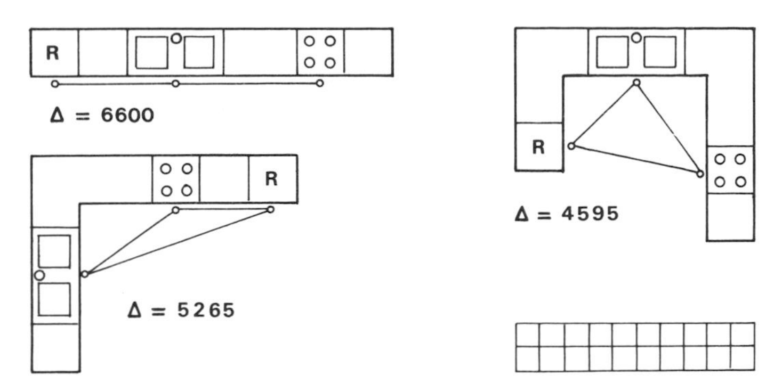

Figure 15.1. *Three kitchen layouts designed, according to the principles discussed in the text, on a 300 mm modular grid. Each includes refrigerator (R), cooker and similar lengths of working surface. The work triangle is indicated and its length (△) given.*

Heights of worktops and fittings

In order to determine an optimal height for kitchen working surfaces, we must consider both the anthropometric diversity of the users and the diversity of tasks to be performed. If, as is commonly the case, a 175 mm deep sink is set into the worksurface, then the effective working level may range from perhaps 100 mm below the worktop height when washing up to a similar distance above it when operating machinery or mixing with a long-handled spoon. We should expect differences, even among tasks performed upon the surface itself, associated with varying requirements of downward force, e.g., between rolling pastry and spreading butter.

Ward and Kirk (1970) studied these matters by means of a fitting trial. The subjects, who were all women, performed three groups of tasks and selected the following worktop heights as optimal:

group A—tasks performed above the worktop (peeling vegetables, beating and whipping in a bowl, slicing bread), 119 [47] mm below elbow height;
group B—tasks performed on the surface (spreading butter, chopping ingredients), 88 [42] mm below elbow height;
group C—tasks involving downward pressure (rolling pastry, ironing), 122 [49] mm below elbow height.

These results were subsequently confirmed using a variety of physiological measurements (Ward 1971) in which it was also shown that the optimal height for the top edge of the sink was approximately 25 mm below the elbow. In order to give some sensible priority to the three groups of tasks studied by Ward and Kirk (1970), I conducted an informal survey of a small number of experienced home-makers who agreed that group B tasks were the most important and group C the least. Giving an arbitrary weighting of 4 to group B and 1 to each of the others, we arrive at an overall optimum of 100 mm below elbow height for the worktop. Combining these with our standard anthropometric data (allowing 25 mm for shoes) gives the figures in Table 15.1.

Table 15.1. Predicted optimal heights (mm) of kitchen equipment.

	Men			Women		
	5th %ile	50th %ile	95th %ile	5th %ile	50th %ile	95th %ile
Worktop	930	1015	1105	855	930	1005
Sink	1005	1090	1180	930	1005	1080

The dimensional co-ordination of kitchen equipment is obviously desirable, and the purchaser should have every confidence that a new oven will fit into the existing range of units. The provision of adjustable worktops, or fittings in a range of heights, is not incompatible with this goal—but it does obviously make things more difficult (and therefore costly). In response (presumably) to Ward's studies, BS 3705 reads thus: "Subject to the need for field research and solving technical problems, it is thought that a 50 mm incremental range of heights of working surfaces may be adopted in the future, the ranges being 900 to 1050 mm for sinks and 850 to 1050 mm for worktops. Because studies show that generally the worktop surface needs to be higher than the present 850 mm for the greater number of users, this standard is omitting the 850 mm worktop height, although this might be included in any subsequent range after the above research is completed. As an interim measure the standard will remain at the BS 3705 (imperial) height, rounded off metrically to 900 mm for sinks and worktops." Department of the Environment (1972) echoes these views, which are quite in accord with the ergonomic evidence—except perhaps that they reinforce sex role stereotypes, being a better match to women's dimensions than to men's.

A decade later the 'interim measure' had acquired a distinct air of permanence and the laudable aspirations had been tempered by experience. BS 6222 reads: "The co-ordinating heights of all units and appliances shall be as follows . . . top of worktop: either 900 or 850 mm (second preference)." Has the ergonomic voice cried out in the economic wilderness? The reader may draw his own conclusions.

A typical set of kitchen units is shown in side elevation in Figure 15.2. The worktop is lower than optimal for approximately half of women and most men. The sink (with its rim at the same height) is lower than optimal for almost everyone. Unfortunately the data do not provide us with tolerance bands so we cannot calculate what proportion of users will be seriously mismatched. Let us postulate a tolerance band of ±50 mm about the optimum. Then a worktop more than 150 mm or less than 50 mm below the elbow is a serious mismatch and the tolerance band for sinks is 75 mm higher. Table 15.2 was calculated upon the basis of this assumption (see Section 2.4 for the method of calculation). The exact figures

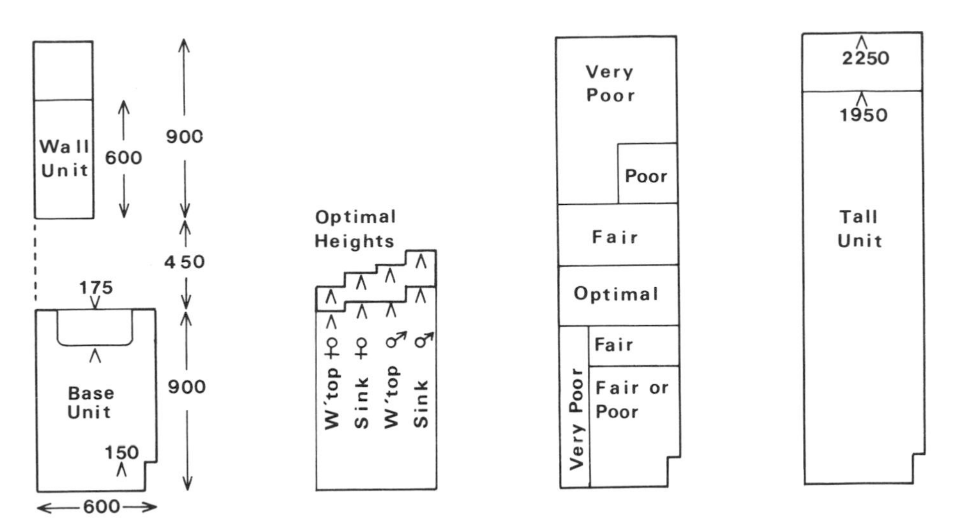

Figure 15.2. Standard kitchen units: (far left) base and wall units and (far right) a tall unit. (centre left) Optimal ranges of worktop heights (5th–95th %ile). (centre right) Analysis of storage space according to the criteria of Table 14.4.

Table 15.2. Percentage of users accommodated by standard 900 mm worktop and sink.

	Too low	Too high	Accommodated	Seriously mismatched
Worktop				
Women	33	4	63	37
Men	55	0	45	55
Sink				
Women	88	0	12	88
Men	100	0	0	100

are, of course, arguable since the tolerance band is arbitrarily set—but the general message is clear.

A comparison of the typical configuration of base and wall unit, with the storage recommendations of Table 14.4, is also revealing. The provision of storage space in the optimal height range of 800–1100 mm is minimal. Tall cupboard units will, of course, ease this problem but limited floor/wall space and the proliferation of whiteware tends to restrict their deployment to the minimum required for brooms and other long objects. Hence, the most accessible storage space in the kitchen is generally the worktop—which rapidly disappears beneath a clutter of foodmixers, spaghetti jars and other homeless objects. The old fashioned walk-in larder is probably the most ergonomically sound solution to kitchen storage requirements—but its integration into the floor plan requires careful consideration.

15.2. The bathroom

The bathroom should combine hedonistic luxury with functional efficiency. It is an environment in which to relax and unwind, soaking in a hot tub, but also a configuration of workstations for the practical activity of washing, grooming and excretion (assuming a special room is not set aside for the latter). *The Bathroom* by Alexander Kira (1976) is a classic of user-centred design research, which every interested person should endeavour to read.

The bathtub

The bathtub presents interesting problems of dimensional optimisation. It must be large enough for comfortable use by one person (or perhaps two) but should not have needless volume, requiring filling with expensively heated water. It is also a notoriously hazardous environment for the frail and infirm.

Two principal postures are adopted in the bath: a reclined sitting position and a recumbent position (possibly with the knees flexed) in which the body is submerged to neck level. For comfort in the sitting position the horizontal bottom of the tub must be sufficient to accommodate buttock–heel length (95th %ile man = 1160 mm) and the end of the bath should provide a suitable backrest. Kira (1976) recommends a rake of 50–65° from the vertical and contouring to conform to the shape of the back. This seems excessive to me—a rake of 30° and a suitable radius where the base meets the end should be quite adequate—we are not particularly looking for postural support since the buoyancy of the water will both unload the spine and lift it away from the backrest. The more we increase the length of the horizontal base the greater the possibility for total submersion. We may shorten our recumbent bodies by around 100 mm by flexing our knees—given that we wish to keep our heads above water—and that as the 95th %ile male shoulder height is 1535 mm there seems little point in lengthening the horizontal part beyond around 1400 mm.

The width of the bath must at least accommodate the maximum body breadth of a single bather (95th %ile man = 580 mm). The ergonomist, who is usually a broadminded sort of person, should also consider the accommodation of couples. Methods for calculating the combined dimensions of more than one person are discussed on pp. 40–41 and Section 11.3. For couples wishing to sit side-by-side the necessary clearance is given by their combined shoulder breadth (920 mm for a 95th %ile couple of the opposite sex). For couples sitting at opposite ends (probably the more common arrangement) the clearance is given by combined breadth of the hips of one person and the feet of the other. This is greatest when the hips are female and the feet are male, in which case the 95th %ile combination is 625 mm. This arrangement does, however, demand that the taps should be in the centre to avoid arguments (Figure 15.3).

Consider a 95th %ile man (sitting shoulder height 645 mm) reclining against the end of the bath. His shoulders will be $645 \cos 30 = 558$ mm above the base of

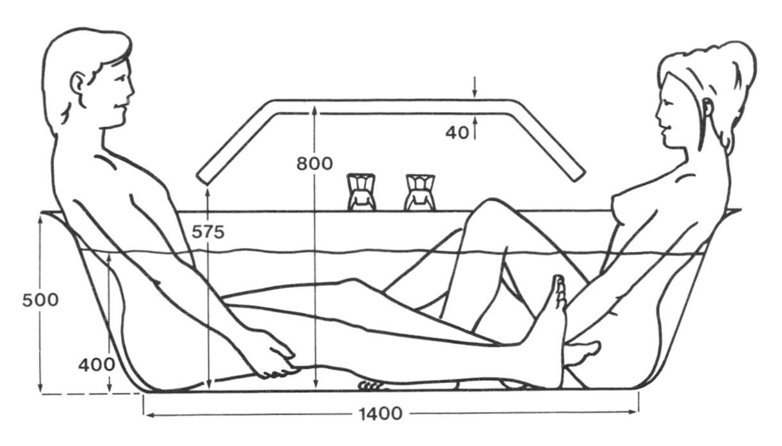

Figure 15.3. The ergonomically designed bath.

the bath. He could not reasonably require more than 400 mm of water. If the backrest was raked further, say to 45°, then 300 mm of water would suffice. (These figures are mere speculation; it would be very interesting to perform a fitting trial to find out what depth of water people really do want). Assuming a 30° rake a bath depth of 500 mm would be required for an adequate quantity of water without too much danger of it splashing over the edge. In fact, a typical tub depth at the present is about 380 mm (15 in), although older models are often deeper. The outside height of the rim (above the floor) is, of course, generally greater than the tub depth (often by as much as 100 mm).

A deep bath or a high rim is generally deemed to make entering and leaving more difficult and hazardous—although Kira (1976) casts some doubt on this, arguing that the manoeuvres people use to enter and leave baths have been insufficiently analysed. Grandjean (1973) cites evidence that a height of 500 mm is acceptable to most elderly or frail people. Grab rails are usually advocated as an aid to stability. These could reasonably be a little above knuckle height at the point where you climb in (e.g., 800 mm above the bath base), around shoulder height (e.g., 575 mm) at the sitting end and about 40 mm in diameter. Vertical grab rails may well be better for the infirm. Additional holds along the side of the bath are also desirable and, for the frail, a nonslip mat inside the bath is essential.

The shower

A minimum floor space of 900 mm × 900 mm is recommended but 900 mm × 1200 mm is preferable (Grandjean 1973). The shower head should be 1900–2000 mm high but preferably adjustable. A seat and grab rail would be advantageous.

The handbasin

This device will be used for washing the hands and face and sometimes the hair. The criteria are relatively simple: it should be possible to wet the hands without water running down the forearms and bending should be minimized. Hence, a basin rim which is at about the elbow height of a short user would be appropriate (5th %ile woman = 930 mm unshod). Kira (1976) studied the above activities experimentally by observing subjects first miming the actions without the constraints of an appliance and then using an adjustable rig. On the basis of these fitting trials he concluded that, for washing the hands, the water source should be located some 100 mm above the rim of the basin, which should be set at 915–965 mm. Conventional handbasins are very much too low (commonly less than 800 mm)—except perhaps for use by children. The present practice of placing the taps at or below the level of the rim seems based on the assumption that people will fill the bowl and wash in the water therein. In fact, according to Kira, 94% of people prefer to wash under a running stream of water.

The water closet

A major controversy exists as to whether the seat of the WC should be at its conventional height (e.g., 400 mm) or very much lower. A substantial body of opinion holds that a squatting position, in which the thighs are pressed against the abdomen, is an aid to the efficient emptying of the bowel—and, hence, will help prevent a variety of unpleasant and potentially life-threatening diseases to which civilized people are prone, largely as a consequence of diet and lifestyle (see, for example, Hornibrook 1934).

McClelland and Ward (1976, 1982) report both an anthropometric survey and

Table 15.3. Anthropometric data relevant to WC design (all dimensions in millimetres; unclothed and unshod).

	Men				Women			
Posture	5th %ile	50th %ile	95th %ile	SD	5th %ile	50th %ile	95th %ile	SD
Sitting								
Seat height[a]	345	410	475	40	290	370	450	50
Perineal length[b]	200	250	300	30	140	190	240	30
Squatting								
Seat height[a]	45	160	275	70	100	180	260	50
Perineal length[b]	240	290	340	30	170	220	270	30
Preferred seat height[c]	380	430	480	30	345	400	455	33

[a]Vertical distance from floor to buttocks.
[b]Horizontal distance from buttock-cleft to genitalia.
[c]As determined in fitting trial.
Data from McClelland and Ward (1976, 1982).

a fitting trial concerned with WC design. In the former, subjects were photographed in three unsupported postures including a 'conventional' sitting position and a squatting position. Results are summarized in Table 15.3. A subsequent fitting trial showed quite clearly that subjects were most comfortable at a conventional range of heights—it could be argued, however, that in these respects we do not necessarily know what is good for us.

Lowering the WC seat to 200 mm or less would cause considerable problems for the infirm who have difficulties in standing up and sitting down, even with conventional appliances. The entire question is at present unresolved.

Mirrors

Bathrooms often contain a small mirror in which men shave and women make up their faces. It has been my experience that these are approximately equally divided into those which are too high for convenience and those which are too low—since I now have a beard the following comments are provided in the spirit of altruism.

Consider a person standing erect and looking at herself in a mirror. Her virtual image is an equal distance behind the mirror as she is in front (Figure 15.4). A line of sight drawn from her eyes to the image of her feet cuts the plane of the mirror at exactly one-half eye height. Similarly, the line of sight to the image of her shoulders cuts the mirror halfway between eye height and shoulder height, etc. The necessary dimensions of mirrors may be calculated on this basis. In Table 15.4

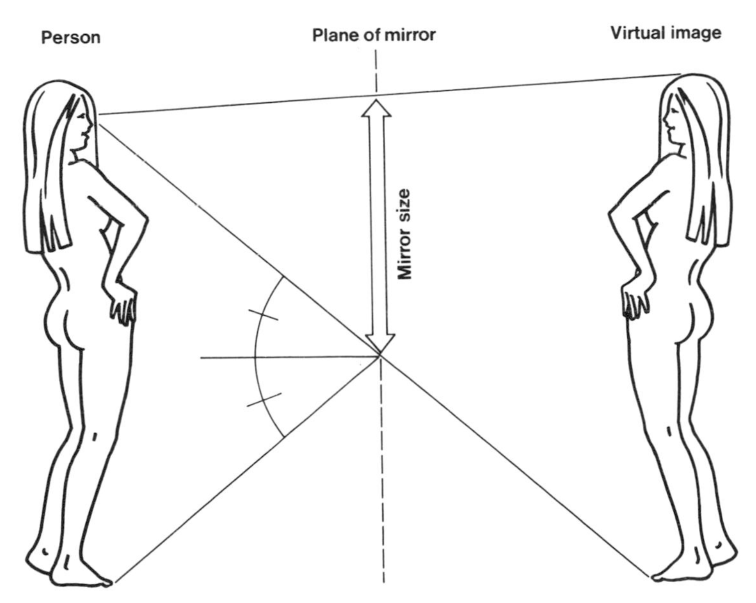

Figure 15.4. Required dimensions for a full-length mirror. Note that the position in which the person stands is of no relevance, since her virtual image is always an equal distance behind the mirror.

Table 15.4. Minimum dimensions of mirrors.

	Height (mm)		Width (mm)
	Top	Bottom	
Face only	1825	1360	200
Whole body	1825	705	600

the top is based on a 95th %ile man and the bottom on a 5th %ile woman. It is interesting to note that this construction, which will be found in almost any elementary textbook of physics, predicts that the distance you stand away from a mirror will not alter how much of your body you can see. This is contrary to experience and I cannot account for the disparity.

15.3. The dining room

Somewhat surprisingly, this essential workstation has received very little ergonomic scrutiny. Dining tables and chairs are, by convention, relatively high—typically 460–480 mm and 760 mm, respectively. Is there any special reason for this? An upright posture presumably aids digestion but we should not forget that the ancients reclined at the *mensa* without apparent discomfort.

A place setting requires 685 mm × 430 mm of table surface, minimum elbow room is 740 mm per diner (Woodson 1981).

15.4. The bedroom

Tall people commonly find standard beds and mattresses too short. Working on the assumption that a length of 150–200 mm greater than stature is required (for pillows, ankle flexion, arm positions, etc.) a mattress length of 2100 mm will accommodate the 99th %ile man. A width of 1000 mm will allow a full range of sleeping postures for a single bed; similarly, 1800 mm is a desirable width for a double bed but begins to cause problems with bed-making. Beds of around 550 mm in height are currently fashionable. This is convenient for getting in and out but is on the low side for convenience in bed-making (Grandjean 1973).

Many people believe that a 'hard' mattress is required to prevent or alleviate back pain. Clinicians tell me of cases in which a patient's back has apparently been made worse on changing to a hard 'orthopaedic' mattress. It may well be that the whole issue is based on a misconception and that the hardness of the mattress *per se* is of little relevance in comparison with the characteristics of its underframe (bedstead). If the latter is excessively elastic or deformed with age, it will allow the entire bed to sink into a hammock shape—which is very probably bad for the back. We might call this latter characteristic 'sag'—and distinguish it from the 'conformability' of the mattress (i.e., its ability to adapt to body contours and distribute pressure evenly). In my opinion there is no real reason for a mattress to be uncomfortably hard in the interests of the spine—it is a firm underframe that is really required. The mattress and bedstead combined should give conformability without sag.

Parsons (1972) published an extensive review of the ergonomics of beds and bedrooms in which he observed, *inter alia*, that there had been no scientific studies of the relationship between sexual behaviour and bed design. I believe this statement still to be true but find it remarkable.

15.5. Miscellaneous architectural features

Stairs, etc.

Change of level may be accomplished in various ways. The choice depends, in the first instance, upon the angle of ascent. MIL-STD-1472C (Department of Defense 1981) recommends:

ramps: 0–20° acceptable, 7–15° preferred;
stairs: 20–50° acceptable, 30–35° preferred;
stair ladders: 50–75° acceptable, 50–60° preferred;
ladders: 75–90° acceptable, 75–85° preferred.

In the domestic context we are only really concerned with stairs (except for cellars, lofts, etc). According to Grandjean (1973) a 30° staircase is physiologically most efficient (in terms of energy expenditure). He therefore recommends risers of 170 mm and treads of 290 mm as preferable for most applications.

Limitations of space will commonly require a steeper stair; Ward and Beadling (1970) recommend, for houses with a floor to floor height of 2600 mm, a 12-riser stair with a riser height of 217 mm and a tread of 245 mm (angle 41°).

Stairways should have at least one and preferably two easily gripped handrails, located 850–900 mm above the front edges of the treads. Wooden rails of circular cross-section 50–70 mm in diameter are very suitable but many other designs would do just as well. Parapets, bannisters, etc., should be at least 1000 mm high to afford reasonable protection from toppling over. It is invidious to specify a minimum separation of rails which will prevent children from getting their heads stuck—since any separation will allow them to get something or other stuck.

Flooring

Non-slip flooring is an important safety feature. Furthermore, a resilient or 'sprung' floor is very much less tiring to walk on than a rigid one—and in the ballroom or dance studio it is essential. Thermal and auditory insulation is also important. Table 15.5 summarizes the slip resistance of some common flooring materials, according to Sigler (1943).

Table 15.5. Slip resistance of common flooring materials.

Material	Wet	Dry
Granite	G[a]	G
Marble	P	G
Concrete	P	G
Ceramic tiles	G	G
Cork	M	G
Linoleum	P	G
Wood	M	G

[a]G = good, M = moderate, P = poor.
According to Sigler (1943).

Chapter 16
Handles and handtools

The purpose of a handle is to facilitate the transmission of force from the musculoskeletal system of the user to the object grasped. In general, to optimise force transmission is to optimise handle design.

Functional anatomists have made several attempts to classify the infinite variety of actions of the hand. The most obvious division is between gripping (prehensile) actions, in which the hand forms a closed kinetic chain which encompasses an object, and non-gripping actions, such as poking, pressing, stroking, slapping, etc. Intermediate between these classes are hand configurations such as the 'hook', by which we carry a heavy suitcase, or the 'scoop', by which we pick up a handful of small objects. Open chain actions are frequently transient in duration.

Napier (1956) divided grips into:

(*i*) power grips—in which the fingers and thumb are used to clamp the object against the palm;
(*ii*) precision grips—in which the object is manipulated between the pads of the fingers and thumb (see Figure 16.1).

This classification has great appeal in its simplicity and it accurately reflects the ways in which people grasp objects, but we should note that in a large number of manual tasks 'power grips' are used precisely.

16.1. Handle design—basic principles

The following basic principles stem more from common sense than scientific investigation. None the less, they are commonly violated.

(*i*) Force is exerted most effectively when hand and handle interact in compression rather than shear. Hence, it is better to exert a thrust perpendicular to the axis of a cylindrical handle than along the axis (F_b in Figure 16.2 rather than F_a). If the latter is necessary a knob on the end will give extra purchase.

(*ii*) All sharp edges or other surface features, which cause pressure hot-spots when gripped, should be eliminated. These include:

(*a*) 'finger shaping' (unless designed with anthropometric factors in mind);

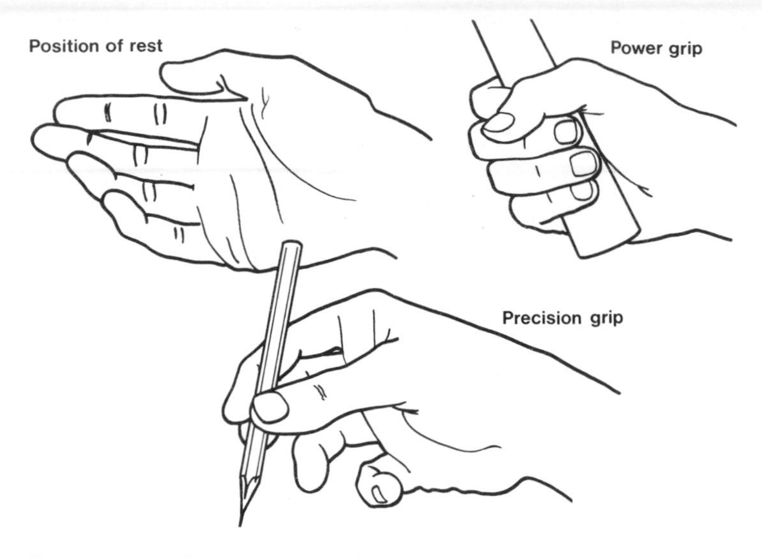

Figure 16.1. *The position of rest, the power grip and the precision grip.*

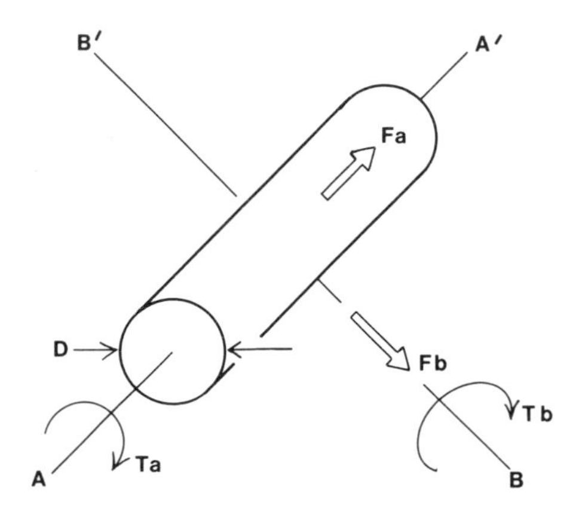

Figure 16.2. *The cylindrical handle showing the long axis* A–A′ *and the perpendicular axis* B–B′.

(*b*) the ends of tools such as pliers, which may dig into the palm (if the handle is short);
(*c*) the edges of flat or raised surfaces, e.g., for the application of labels, logos, etc.;
(*d*) 'pinch points' between moving parts such as triggers, etc.

(*iii*) Handles of circular cross-section (and appropriate diameter, e.g., 30–50 mm) will be most comfortable to grip since there will be no possibility of hot spots—but they may not provide adequate purchase. Rectangular or polyhedral sections will give greater purchase but will be less comfortable. In general, wherever two planes meet (within the area that engages the hand) the edges should be rounded; there are no exact figures but a minimum radius of curvature of about 25 mm seems reasonable.

(*iv*) Surface quality should neither be so smooth as to be slippery nor be so rough as to be abrasive. The frictional properties of the 'hand/handle interface' are complex since the skin is both visco-elastically deformable and lubricated. Heavily varnished wooden handles give a better purchase than metal or plastic of similar smoothness. The explanation is possibly in their resilience (elastic compliance). Rubber is similar but becomes 'tacky'. The subject is worthy of investigation.

(*v*) If part of the hand is to pass through an aperture (as in a suitcase or teacup) adequate clearance must be given. It is remarkable how often this perfectly obvious principle is violated. The following spaces will accommodate virtually all users with a slight leeway:

For the palm, as far as the web of the thumb (as in the handle of a suitcase), allow a rectangle 110 mm × 45 mm.

For a finger or thumb, a circle 30 mm in diameter will allow insertion, rotation and extraction.

(*vi*) When the hand is in its 'resting position' (Figure 16.1), relaxed but ready for action, the fingers are slightly spread and somewhat flexed. The index finger is the least flexed and the little finger the most—hence the pads of the fingers could be arranged approximately around the circumference of a circle. Very roughly speaking they occupy a 60° arc of a circle of diameter 125–175 mm. In gripping devices, such as pliers, the concave surface described by the fingers opposes a convex surface formed by the heel of the palm and base of the thumb. Handles should, if possible, reflect these two curves.

(*vii*) Approximately 10% of the population are left handed.

16.2. *Biomechanics of hand–handle interactions*

Wrist posture

It is highly desirable that the handle of a tool should maintain the neutral position of the wrist (see Section 8.5). This is particularly important if actions of repetitive rotation are to be performed. In the neutral position the axis of a rod gripped firmly in the hand makes an angle of 100–110° with the axis of the forearm.

Gripping

An important group of cutting and crushing tools, from pliers and wire-cutters to nut-crackers and secateurs, are operated by a forceful squeezing action across two pivoting arms. The fingers curl around one arm and the heel of the palm butts against the other. The effective cutting/crushing force is determined by the mechanical advantage of the tool and the user's grip strength. The latter is determined *inter alia* by the distance across the two arms—as shown in Figure 16.3. The optimal handle separation is 45–55 mm for both men and women.

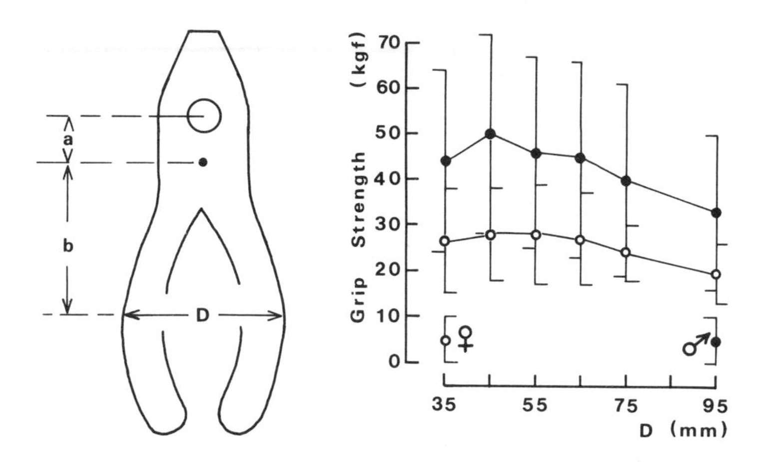

Figure 16.3. Grip strength (G) *a a function of the handle size* (D). *Vertical lines are 5th–95th %ile values in samples of 22 men and 22 women. The tool is a lever of the first class, the mechanical advantage* = b/a. *Hence, the effective cutting or crushing force* = GB/A.
Data from Pheasant and Scriven (1983).

Gripping and turning

Consider a cylindrical handle as shown in Figure 16.2. It may be gripped and turned about its own axis A–A' or about a perpendicular axis B–B'. Screwdrivers employ the former action; T-wrenches the latter.

When the handle is employed as a T-wrench, the available torque (T_b) is determined by the angle–torque curves for the actions of supination and pronation (see Figure 8.3). (The former movement is equivalent to a clockwise turn with the right hand or an anticlockwise with the left; the latter is the opposite.) For rotation about axis B–B' the available torque (T_b) is, within reasonable limits, independent of the design of the handle.

When the handle is employed as a screwdriver (rotating about axis A–A') the strength of the action is no longer determined by the user's capacity to generate torque but by the ability to transmit it across the hand–handle interface. It is, therefore, strongly dependent upon handle design. Torque about axis A–A' is exerted by a shearing (frictional) action on the cylinder's surface, hence,

$$T_a = G\mu D \qquad (16.1)$$

where G is the net compressive force (i.e., grip), D is the diameter of the cylinder and μ is the coefficient of limiting friction between the hand and the handle. For any handle of circular cross-section (i.e., cylinder, sphere or disc), T_a will increase with diameter. We should also expect the grip strength (G) to be dependent upon diameter (the optimum value of which can only be determined empirically). Figure 16.4 summarizes the results of such an experiment. Few 'real' handles are

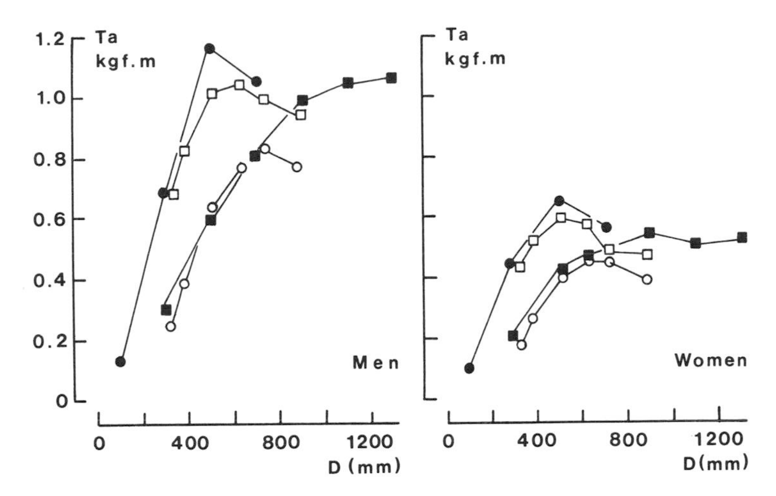

Figure 16.4. Torque (T_a) *as a function of handle diameter (D). Average of clockwise and anti-clockwise exertions. Note that the relative strengths on using the various handles are similar for women and men.*
● = *knurled steel cylinders;* □ = *polished wooden cylinders;* ○ = *polished wooden spheres;* ■ = *knurled discs.*

actually circular in cross-section—but quite substantial irregularities in shape seem to make surprisingly little difference. Hence, commercially available screwdrivers (London pattern, cabinet makers', engineers, etc.) perform no better in these tests than do knurled steel cylinders of equivalent diameter (Pheasant and O'Neill 1975, Pheasant and Scriven 1983). Subsequent (unpublished) experiments have shown that the same is true for a variety of devices such as taps and doorknobs. However, torques exerted about axis *B–B'*, as in using T-shaped or L-shaped devices, are very much greater. (The torque which may be exerted using a typical L-shaped lever-type door handle is in the order of twice that available from any cylinder, sphere or disc turning about its own axis.)

The strength of a thrusting action along axis *A–A'* is given by

$$F_a = G\mu \qquad (16.2)$$

The diameter is only relevant as a determinant of G. Hence, we find that the optimal diameter for axial thrusts is somewhat less than that for turning actions. Data are summarized in Table 16.1. It is also worth noting that the maximal hand–handle contact area occurs on handles 50–60 mm in diameter—which will,

Table 16.1. Handle sizes which allow the greatest force/torque in operation.

Pivoting tools	
Distance across arms (mm)	45–55
Handles of circular cross-section	Diameter (mm)
Cylinders	
Axial thrust (F_a)	30–50
Axial rotation (T_a)	50–65
Spheres	
Axial rotation	65–75
Discs	
Axial rotation	90–130

For cylindrical handles used to exert force or torque perpendicular to the axis (F_b, T_b) the diameter is not critical; a diameter of 30–50 mm is suitable.

therefore, minimize the surface stress to the skin (Pheasant and O'Neill 1975, Grieve and Pheasant 1982).

Sex differences

An increasing number of women are entering traditionally 'male' occupations involving the manual working of wood, metal, stone, etc. Since women have smaller hands and lesser strength than men, do they require differently designed handtools? Ducharme (1975, 1977) surveyed the experiences of female engineers and concluded that "women workers find male tools unsatisfactory" (*sic*). For example, the arms of pivoting tools were deemed to be too widely spread for the small female hand. However, Figures 16.3 and 16.4 reveal that, although women are weaker in all conditions tested, the relative strengths in different conditions are similar for both sexes—notwithstanding the difference in hand size. Anthropometrically, this is not as surprising as it seems. The average male and female hands are 190 and 175 mm in length, respectively, but in gripping actions the hand is flexed. Let us suppose that the optimal handle size for an individual is one-third hand length. The overall male and female optima would differ by only 5 mm—which would be masked in most experimental studies and of little practical consequence anyway. There is little evidence to suggest that women require a special feminine toolkit—rather it seems that men have reserves of strength which enable them to overcome problems associated with the suboptimal design which women find insuperable. For example: commercial screwdrivers rarely have an effective handle diameter of much in excess of 30 mm; plier/cutters typically have a mechanical advantage in the order of only 6 : 1; and the wheel braces of motorcars often provide totally inadequate leverage (Pheasant and Scriven 1983). In each case the re-design of the tool would benefit both male and female users.

The length of the screwdriver shaft

Most carpenters and other craftsmen will tell you that the longer the shaft of a screwdriver the greater the torque. There is presumably some factual basis for such a widespread belief—but simple mechanics would not predict such a relationship and the author has repeatedly failed to demonstrate it in the laboratory (albeit using a static test rig). The phenomenon, which, if it exists at all, only does so under dynamic conditions, is something of a mystery. Hypothetical explanations include the aiming of the screwdriver blade in the slot and the torsional elasticity of the shaft.

Chapter 17
Controls and displays

An excessive concern for such design minutiae as the dimensions of a push button or the scale markings of a meter may be interpreted by some as evidence of an obsessional mental disturbance. None the less, an inadequate consideration of the design of the 'knobs and dials' on a control panel may have serious consequences for the performance of the operator and the safety of everybody concerned. The most notorious example is the three-pointer altimeter, once common in aircraft cockpits, which was subject to a particularly high rate of reading errors (see, for example, Oborne 1981). The disasters which sometimes result from the misreading of a meter are usually attributed to 'human error'—with the clear implication that the operator is to blame. In the ergonomist's view such an attribution is a considerable oversimplification and it would be both more realistic and more helpful, for the prevention of future similar occurrences, to direct attention to such design features as might predispose the operator to err. The purpose of this final chapter, therefore, is to summarize the recommendations of the ergonomics literature for the design of the most common types of controls and displays. It is not always possible to point to a single definitive empirical study from which specific design specifications may be derived—commonly they are repeated from one source to the next with minor modifications. The recommendations which follow have been compiled with reference to Woodson and Conover (1964), Murrel (1969), McCormick (1970), Dreyfuss (1971), HFDNE (1971), Van Cott and Kinkade (1972), Department of Defense (1981), Diffrient *et al.* (1974, 1981), Grandjean (1981) and Woodson (1981). No fixed procedure has been adopted in the attempt to establish a consensus view but, in general, the 'preferred' values are within the limits quoted by most of the above authorities, whereas the maxima and minima may be more contentious.

17.1. Traditional controls

Controls may be divided into two main categories: (*a*) small devices which control electrical or electronic systems by means of switches or potentiometers, which have minimal mechanical resistance, or (*b*) larger items which operate mechanical or hydraulic linkages against substantial resistance. They may be further subdivided according to whether their function is discrete (to activate, switch or select from alternatives) or continuous (to set, adjust, control or track) (Table 17.1). Data entry devices (keyboards, keypads, etc.), which, due to the increasing use of digital circuitry, are tending to take over the functions of the more traditional controls, will be considered separately.

Table 17.1. Characteristics of control devices.

	Pushbutton	Footswitch	Toggle switch	Rocker switch	Rotary selector	Knob	Slider	Joystick	Thumbwheel	Crank	Handwheel	Lever	Pedal (pivot)	Pedal (thrust)
Discrete														
Activate (on/off)	G	G	VG	VG	P	*	P	*	*	*	*	*	*	*
Select three states	A	*	F	P	VG	*	*	*	*	*	*	G	*	*
Select multistate	A	*	P	*	VG	*	VG	*	F	*	*	G	*	*
Continuous														
Set/adjust	*	*	*	*	*	VG	G	*	F	VG	G	G	*	*
Control/track	*	*	*	*	*	P	*	VG	*	G	VG	F	G	*
Exert force	*	*	*	*	*	*	*	*	*	G	G	VG	F	VG
Speed of operation	VG	F	VG	VG	F	F	G	G	P	F	G	G	F	G
Inherent visual feedback	P	*	VG	F	F	F	VG	*	*	*	G	G	*	*

VG = very good, G = good, F = fair, P = poor, A = appropriate in arrays, * = inappropriate for the purpose.

Pushbutton (Table 17.2)

The pushbutton should have a positive snap-action which gives tactile feedback and, preferably, an auditory click; for enhanced feedback an associated indicator or internal light is helpful. A high-friction or dished surface aids finger control—the latter should be matt to prevent annoying reflections. Palm-operated pushbuttons (e.g., for an isolated emergency stop) are commonly domed or mushroom shaped.

Capacitance and membrane touch switches have no moving parts and are easily sealed for cleaning, etc. They have minimal tactile quality and so visual feedback is essential. Auditory feedback in the form of 'bleeps' can be maddening and should be avoided.

Footswitch (Table 17.3)

May serve as an accessory control when the hands are overloaded.

Table 17.2. Pushbutton.[a]

	Minimum	Preferred	Maximum
Fingertip operation			
Diameter (edgelength if rectangular)	6	12–15	25
Travel	3	5–10	35
Resistance	0·15	0·25–0·35	1
Separation			
Single finger, random operation	15	50	—
Single finger, sequential operation	6	25	—
Several fingers	—	12–15	—
For blind reaching	—	100–150	—
Palm operation			
Diameter	20	30–40	—

[a]All dimensions are given in millimetres and resistances in kilogram force. Separations between controls are measured from edge to edge. Displacements are in millimetres unless otherwise specified. The reistance of a rotary control is given in terms of a tangential force applied to its surface or edge. Hence the operating torque is the product of this force multiplied by the effective radius at which it acts about the axis of rotation.

Table 17.3. Footswitch.[a]

	Minimum	Preferred	Maximum
Diameter	13	50–80	—
Displacement	15	20–30	60
Resistance	3	5	9
Separation	100	150	—

[a]See note to Table 17.2.

Toggle or rocker switch (Tables 17.4 and 17.5)

The preferred control for on/off or other two-state selection. May be used in three positions (e.g., off/standby/on) but the rotary selector is preferable in such cases. May be positive snap-action or spring-loaded for momentary action. A variety of toggle shapes are equally satisfactory. If several switches are to be used they should be mounted in a horizontal row rather than a vertical row, since the necessary separation is less. Note: The population stereotype (see Section 17.5) is a major difficulty.

Rotary selector (Table 17.6)

The preferred control for between more than two and less than 24 settings. May be moving pointer/fixed scale or moving scale/fixed pointer. The former case,

Table 17.4. Toggle switch.[a]

	Minimum	Preferred	Maximum
Diameter (or breadth) of tip (or paddle)	3		25
Length	15		50
Number of settings		2	3
Displacement between adjacent settings (deg.)			
Two settings	30	40–60	120
Three settings	18	30	60
Resistance	0·25	0·5–1	1·5
Separation			
Horizontal, random operation		50	
Horizontal, sequential operation	15	25	
Vertical (tips in closest position)	20	50	

[a]See note to 17.2.

Table 17.5 Rocker switch.[a]

	Minimum	Preferred	Maximum
Rocker length	12	25–30	50
Rocker width	6	12–15	25
Displacement (deg.)		30	
Resistance	0·15	0·25–0·5	1·0
Separation			
Random operation	20	50	
Sequential operation	15	25	

[a]See note to 17.2.

Table 17.6. Rotary selector.[a]

	Knob			Pointer		
	Min.	Pref.	Max.	Min.	Pref.	Max.
Diameter	10	15–30	50	na	na	na
Length	na[b]	na	na	25	25–30	100
Width	na	na	na	—	—	25
Depth	15	15–25	75	15	15–25	75
Resistance	0·4		1·3	0·4		1·3
Number of settings	3		24	3		24
Displacement between settings (deg.)	15	30	40	15	30	40
Separation	25	50		25	50	

[a]See note to Table 17.2.
[b]Not applicable.
The resistance of a rotary control is given in terms of a tangential force applied to its surface or edge. Hence the operating torque is the product of this force multiplied by the effective radius at which it acts about the axis of rotation.

generally with a bar-shaped knob, is preferred in most applications. The grasping surface for round knobs should be serrated or knurled. Setting labels should not be obscured by the hand and parallax should be minimized. Scale ends should be clearly demonstrated.

Knobs (Table 17.7)

The standard device for making settings on a continuous scale. A variety of shapes may be appropriate. A folding crank lever may be an advantage for multiturn knobs. Knob diameters have little effect on speed and accuracy provided depth is adequate. See rotary selector for further comments.

Table 17.7. Knobs (continuous adjustment).[a]

	Minimum	Preferred	Maximum
Diameter	10	15–30	80
Depth	15		25
Resistance			0·13

[a]See note to Table 17.2.
The resistance of a rotary control is given in terms of a tangential force applied to its surface or edge. Hence the operating torque is the product of this force multiplied by the effective radius at which it acts about the axis of rotation.

Slider (Table 17.8)

Continuous or discrete, the slider is an exceedingly useful control in that it provides excellent visual feedback—especially good in rows for comparative readings. Stereotypes are problematical (see Section 17.5).

Table 17.8. Slider (continuous or discrete).[a]

	Minimum	Preferred	Maximum
Slide width	10	15–20	20
Slide height	10		
Displacement (between adjacent settings if discrete)	10	15–25	25
Separation			
Random operation	20	50	
Sequential operation	15	25	

[a]See note to Table 17.2.

Thumbwheel (Table 17.9)

Continuous or discrete, it should only be used where shortage of space makes it essential. Surface should be knurled for grip. Stereotypes are problematic (see Section 17.5).

Table 17.9. Thumbwheels (discrete or continuous).[a]

	Minimum	Preferred	Maximum
Diameter	40		65
Width	4		15
Protrusion from surface	3		6
Resistance	0·2		0·5
Separation	6	10	

[a]See note to Table 17.2.

Crank (Table 17.10)

The preferred method of setting mechanical linkages (rack and pinion, etc). The crank handle should be knurled or fluted and rotate freely on its shaft. The available torque increases with the radius of the crank and, in general, the maximal speed of rotation decreases with loading; but when the loading is low a small crank can be turned more rapidly than a large one. Performance is optimized when the axis of the crank is horizontal and makes an angle of 60–70° to the coronal (frontal) plane of the body.

Table 17.10. Crank.[a]

	Small (wrist and finger movements)			Large (hand and arm movements)		
	Min.	Pref.	Max.	Min.	Pref.	Max.
Handle diameter		13			25	
Handle length		40		100	100	
Crank radius	13		120	120		200
Resistance	0·9		2	2	2·4	6
Separation	70	100	na	70	100	na
Top speed (r.p.m.)			270			175

[a]See note to Table 17.2.
The resistance of a rotary control is given in terms of a tangential force applied to its surface or edge. Hence the operating torque is the product of this force multiplied by the effective radius at which it acts about the axis of rotation.

Handwheel (Table 17.11)

Designed for two-handed operation, greater torques may be exerted than with cranks or knobs. Most effective when displacement does not exceed ± 60° from the null position, since the hand position need not be changed (see Section 13.1).

Table 17.11. Handwheel.[a]

	Minimum	Preferred	Maximum
General			
Wheel diameter	200		550
Grip (rim) diameter	20		50
Displacement		90–120	
Resistance	2		13
One hand	2		13
Two hands	2		25
Steering Wheels			
Wheel diameter	350	380–440	500
Grip (rim) diameter		20–40	
Orientation (column to horizontal)	30	50–60	60

[a]See note to Table 17.2.
The resistance of a rotary control is given in terms of a tangential force applied to its surface or edge. Hence the operating torque is the product of this force multiplied by the effective radius at which it acts about the axis of rotation.

Lever (Table 17.12)

Suitable for continuous or discrete control when substantial force is required. When the lever pivots through less than 30° the handgrip should be cylindrical; otherwise spherical.

Table 17.12. Lever.[a]

	Spherical handle			Cylindrical handle		
	Min.	Pref.	Max.	Min.	Pref.	Max.
Handle diameter (sphere)	20	30–50		na	na	na
Handle diameter (cylinder)	na	na	na	20	30–40	50
Handle length (cylinder)	na	na	na	100	100	na
Displacement (minimum between settings, maximum total)						
Forward/backward	50		350	50	100–200	350
Lateral	50		950	50	80–150	950
Resistance						
Push/pull	0·4		1	1		14
Up/down	0·4		1	1		9
Sideways	0·1		1	1		7
Separation	50	100	na	50	100	na

[a]See note to Table 17.2.

Pedals (Table 17.13)

There are two basic types of pedal.

(*a*) For relatively light applications and continuous control (e.g., accelerator of car) operated principally by movement of ankle flexion. These may support the entire foot and be pivoted behind the heel, or be suspended, in which case the heel rests on the floor. The effective radius of this movement ranges from around 160–210 mm.

In a motor vehicle the leg might reasonably be at 55° to the horizontal, so the pedal surface should be at 35° if the ankle is at a right angle, and the centre of the pedal should be around 105 mm above the floor. (About one-third of the way through the pedal's total travel for the neutral position.)

(*b*) For heavier thrusts using the whole leg (e.g., brake of car). The line of thrust should be in the direction of the operator's hip joint.

For a further discussion of pedal location and operating posture see Section 13.1.

Table 17.13. Pedals.[a]

	Light (ankle pivot)			Heavy (leg thrust)		
	Min.	Pref.	Max.	Min.	Pref.	Max.
Length						
Supporting ball only	75	90–110		75	90–110	
Supporting whole foot	225	250–300			250–300	
Width	75	90–110		75	90–110	
Travel (mm)	25		60	25		180
(deg.)		15	30	na	na	na
Resistance						
Foot resting off pedal	1·5		5·5	1·5	6–9	?
Foot resting on pedal	3		5·5	4·5	6–9	?
Separation						
Random operation	100	150		100	150	
Sequential operation	50	100		50	100	

[a]See note to Table 17.2.

17.2. *Data entry devices*

Keyboards and keypads

Postural aspects of keyboard operation have been discussed in Section 8.6 and recommendations concerning the physical characteristics of keyboards are presented in Section 12.4. The patterns in which the keys are arranged, both in the alphanumeric keyboard and the numeric keypad, require further comment —especially since they illustrate the widespread problem of the *de facto* standard and its relation to ergonomic considerations.

The conventional typewriter keyboard, known by the letters of its top row QWERTYUIOP, was devised as long ago as 1874, by Sholes and his colleagues, with the aim of minimising the frequency of jamming on the primitive mechanical devices of the day. It has achieved worldwide acceptance and is incorporated into numerous standards (e.g., BS 5959, ISO 4169). The arrangement is, however, by no means an optimal one—the frequency of occurrence of the various letter combinations in the English language results in an overuse of the left hand and little fingers, and only a relatively small proportion of common words (around 45%) can be typed without moving from the home row. The most notable attempt at solving these problems was that of August Dvorak in the 1930s, who placed the vowels and most commonly used consonants (AOEUIDHTNS) in the home row and allocated a higher portion of the workload to the right hand and stronger fingers. Dvorak's arrangement has a home row allocation of 67% and trials showed that it enabled beginners to learn more quickly, achieve higher speeds and make fewer errors. Why then has it been ignored for half a century? The 'negative transfer' experienced by typists trained on QWERTYUIOP, and the massive problems of retraining which this engenders, effectively dooms to failure all attempts at the design of a more rational layout. Consequently, the overwhelming majority of ergonomists are, on pragmatic grounds, in favour of QWERTYUIOP—to the extent that it could be described as 'the unergonomic arrangement which most ergonomists recommend'.

The following two arrangements seem to have evolved concurrently and independently in the case of numeric keypads:

```
For telephones          For calculators
   1  2  3                 7  8  9
   4  5  6                 4  5  6
   7  8  9                 1  2  3
      0                       0
```

Computers usually (but not always) adopt the calculator layout. It is not easy to see why anyone should have thought of this arrangement in the first place but it seems to work well enough in practice. Conrad and Hull (1968) tested both arrangements and found that naive subjects were slightly faster using the telephone layout and substantially more accurate. When subjects alternated between the two layouts in a single test, their speed was less than for either used separately but their error rates were intermediate. Since both keyphones and computers are becoming increasingly common, the two arrangements must frequently appear side by side on the same desk. The whole situation is in need of rationalisation—preferably in favour of the rational layout.

Alternative devices

A number of alternatives to the keyboard exist which are said to make computers more acceptable to the non-specialist. Regrettably, ergonomic recommendations concerning these devices scarcely extend beyond the self-evident (SGO). Touch screens and light pens have the great advantage of providing a direct and self-evident control/display relationship. They are fatiguing to use for long periods (unless wrist support is provided) and tend to result in unsatisfactory neck postures. Some methods of making the screen touch sensitive result in degradation of the visual image and this becomes more marked with time as the screen inevitably becomes scratched and grimy. A number of special devices share the basic function of moving a cursor around on a screen: the minature fingertip operated joystick; the trackerball, more exotically known as the 'joyball', which is a freely moving sphere in a special mounting; the mouse, which is a small object moving on two perpendicular potentiometric wheels about any available surface. Their relative merits have not been established at present.

It is difficult to estimate the rate at which new technologies, such as voice input, will render all of these devices obsolete. Current speech-recognition devices have a limited vocabulary and need to be individually 'trained' by each operator who speaks to the machine. One exercise in technological forecasting placed the median date for a voice-operated typewriter at 1991. The problem is said to be easier in Japanese than in English (Michie and Johnston 1984).

17.3. *Visual displays*

The cardinal principle in the design of visual displays is to establish precisely what information the user requires in his task and to display no more or no less than this in a form which is visible, legible and intelligible: that is, can be seen, can be read and can be understood.

Visibility is mainly determined by display location. Postural aspects and visual fatigue are discussed in Section 8.4, where it was observed that detailed information can only be acquired from a small central (foveal) portion of the visual field surrounding the line of gaze. Only moving objects or bold contrasts will be observed in the remaining peripheral field.

Legibility is determined by visual acuity—that is, the capacity of the eye to resolve fine details. Under adequate viewing conditions people with normal visual acuity will be able to resolve features which subtend an angle of 1′ of arc at the eye. In general, legibility will increase with:

(*i*) the ratio of the size of the features of the display to the distance at which it is viewed;
(*ii*) contrast between features and background;
(*iii*) overall illumination;

(*iv*) certain aspects of the visual form of the object, such as typographical characteristics;

(*v*) the length of time for which the display is presented to the observer (duration);

(*vi*) the familiarity of the 'message'.

Intelligibility is determined by the meaning that the display has for the observer. A display message may be legible but meaningless. Such messages are known, both to consumerists and computer programmers, as 'gobbledegook'. Achieving intelligibility is widely considered to be an art but it has received some scientific scrutiny, particularly with respect to instructional text and bureaucratic communication (Wright and Barnard 1975, Hartley and Burnhill 1977).

Alphanumeric information

The following pertain to such visual media as labels on packages, products and equipment; to posters, notices and signs; and in some measure to printed text. It will be assumed throughout that legibility is of paramount importance. To some extent the factors contributing to legibility (listed above) may be traded-off against each other—the more one is enhanced the greater the leeway with the others. Hence, if letters much larger than the minimum are used on a well-illuminated poster which will be viewed at leisure, the designer has considerable freedom to adopt an unconventional typeface which would be quite unacceptable under less fortunate circumstances.

Size of letters and digits

For persons with normal vision, viewing optimally designed text in good indoor lighting, the letter-height must exceed $0{\cdot}0028D$, where D is the reading distance (BS 3693). In less favourable circumstances a letter height of up to $0{\cdot}01D$ may be required. In general, the larger the letters the more visible they become, up to the point (ill-defined) at which the message begins to 'break up' as the demand for additional scanning impairs performance. For most purposes a letter height of between $0{\cdot}005D$ and $0{\cdot}0075D$ may be considered optimal. Note that typical values for reading distance D are 400 mm for a book or 720 mm for a display panel with controls which are operated at arm's length.

Letter style and typeface (fount)

Most commonly encountered typefaces are legible under reasonable reading conditions but some are more legible than others and the differences may be critical under adverse circumstances. They have, therefore, been extensively researched in the military context. The results, as presented in McCormick (1970), may be summarized thus:

Capital letters such as B or E have, in their basic block forms, five vertical features and three horizontal features. Assuming each of these to be equally important for recognition, they should each subtend an equal visual angle at the eye, hence the width (W) and height (H) of the characters should stand in the ratio 3 : 5 ($W = 0{\cdot}6H$). Empirical studies have shown that such letters are a little too narrow, and for optimum legibility a character width of between $0{\cdot}7H$ and $0{\cdot}85H$ is optimal for daylight conditions (but for transilluminated displays W should be equal to H). Furthermore, a stroke width of between $0{\cdot}125H$ and $0{\cdot}133H$ is optimal for black figures on a white background whereas for white figures on a black background between $0{\cdot}1H$ and $0{\cdot}125H$ is better. For poor illumination or a very great viewing distance a stroke width of $0{\cdot}2H$ may be required.

For short, attention-getting messages, plain sanserif founts are preferred and in the interests of clarity uneven stroke widths, extended serifs and internal stripes should be avoided. Gill Sans, Futura Medium, Univers, Granby Bold and Helvetica Medium are all highly suitable (indeed, they are virtually indistinguishable to the untrained eye). When departing from these well-tried alternatives it is prudent to check that the following commonly confused character combinations are readily distinguishable: EF, B8, GC, OQD, Z2, 35, 1I, UVY (Woodson 1981).

Typographers generally believe that continuous text is most legible when set in a not-too-extreme serif face, such as Times Roman, Garamond, Baskerville or Plantin (K. Morton 1985, personal communication). The figures in the present book are labelled in Univers and Helvetica Medium and the text is set in Bembo.

Capitals or lower case?

The relative legibility of capitals only versus capitals and lower case is surprisingly contentious. McCormick (1970) and Oborne (1981) both review the subject and cite evidence that printed text (including, for example, newspaper headlines) is more easily read when presented in mixed capitals and lower case. The reason is considered to be that the ascenders and descenders lend words more distinctive shapes. This factor is specifically exploited in typefaces such as Frutiger in which the tops of capitals are slightly lower than the lower-case ascenders. However, capitals alone are considered preferable for the labels on instrument panels (McCormick 1970, Van Cott and Kinkade 1972, Department of Defense 1981). Road signs sometimes use capitals alone and are sometimes mixed. The weight of evidence probably falls in favour of mixed but there may be cases in which capitals allow a more efficient utilization of limited space, hence giving larger letters and better legibility.

Spacing and line length

On average, a spacing of one stroke width ($\pm \frac{1}{2}$) between characters and three stroke widths ($\pm \frac{1}{2}$) between words is about right (Woodson 1981). However, it is good typographical practice (and ergonomically sound) to vary this for different letter combinations, thus aiding the reading of words as units. Hence in the word THE, the T and H are set closer together than the H and E. The spacing between lines should not be less than $0{\cdot}33H$ and there should be a gap of at least one stroke width between the ascenders of one line and the descenders of the next.

When space is limited it is better to use a narrow typeface (such as Helvetica Bold Condensed) than to squeeze optimally proportioned letters together.

Experience in the layout of successful advertisements suggests that seven words per line (or 42 characters including spaces) is easiest for the reader to scan (K. Morton 1985, personal communication).

Instrumental displays

Analogue or digital displays?

Digital displays are ergonomically preferable in most situations in which the user must make an accurate quantitative reading. (Furthermore, the cheapness and availability of liquid crystal displays (LCDs) and light emitting diode (LED) displays, and the ease with which they may be employed in digital circuits, make them technically preferable also.) However, traditional analogue dials have distinct advantages in the following circumstances:

(*a*) Where the readings change rapidly. Digital displays are almost unreadable or at best difficult to take in (depending on the sample rate). It is easier to make a visual estimation of the average position of a moving pointer than to mentally average a set of numbers. Digital displays should only be used if all but the last digit (of a three or more figure reading) are unlikely to change more than once every 5 s.

(*b*) Where not only the instantaneous value but also its rate of change are important, moving pointer scales give a better indication.

(*c*) Where indications of relationship are important, e.g., where two displays must be read and compared with each other or where there are certain preset critical or limiting values. In such cases it may be advantageous to design a dial which conveys the necessary information and no more, e.g., 'too cold, OK, too hot', rather than using a quantitative display at all.

To summarize, it is generally considered that the circular analogue dial encodes and communicates information in visual and geometric form as well as numerically, but that the actual figures it presents (if relevant) are usually more difficult to acquire. It is perhaps surprising therefore that Galer *et al.* (1980) discovered digital speedometers for motor-cars to be satisfactory and probably actually preferable to analogue ones. If more examples of this kind are forthcoming perhaps, as digital displays become more familiar to the unsophisticated user, the traditional view outlined above will require revision.

The design of analogue displays

The evidence cited in the standard ergonomic reference works allows the following conclusions.

(*a*) An 'open window' display on which a pointer moves along or around a scale is preferable to one in which the scale moves past a window or pointer.

(*b*) A circular scale is preferable to a linear one.

(*c*) A horizontal linear scale is preferable to a vertical one since the motion stereotype (see Section 17.5) is less ambiguous—except when a row of dials are used for comparative readings in which case the vertical is better.

(*d*) The smallest divisions marked on a scale should never be less than the reading accuracy which is required for successful performance of the task at hand. Subdivisions should be into two or five parts, leading to graduations which progress in multiples of five or two, respectively. Progression by threes, fours, sixes, etc., are to be avoided. (Skilled users are generally able to interpolate the space between markings into five parts.)

(*e*) Grandjean (1981) recommends the following minimum dimensions of scale markings where D is the viewing distance.

Length of major graduations	$D/90$	$(0{\cdot}011D)$
Length of intermediate graduations	$D/125$	$(0{\cdot}008D)$
Length of minor graduations	$D/200$	$(0{\cdot}005D)$
Thickness of graduations	$D/5000$	$(0{\cdot}0002D)$
Distance between minor graduations	$D/600$	$(0{\cdot}0016D)$
Distance between major graduations	$D/50$	$(0{\cdot}02D)$

These recommendations lead to a scale which is around 2·5–3 times bolder than one designed according to BS 3693 or Murrel (1969). BS standard scales have a nominal reading distance given by $D = 14{\cdot}4L$ (where L is the scale base length between its terminal marks). A standard scale may therefore be expected to meet Grandjean's boldness criteria at a reading distance of $D = 6L$.

(*f*) The pointer (index) should obscure neither the scale nor the numbers. Ideally, it should be in the same plane as both to avoid parallax errors in reading.

CRT displays

The layout of VDT workstations is discussed in Section 12.2. The following recommendations for the characteristics of the screen itself are based on Cakir *et al.* (1980), Eriksson (1983) and Health and Safety Executive (1983).

Polarity

In genereal, dark images on a light screen (positive polarity) are preferable to the opposite (negative polarity) although the latter are much more commonly encountered in practice (Radl (1980) and Bauer and Cavonius (1980) showed that performance was better; furthermore, the positive screen is more compatible with ordinary office lighting levels and less subject to reflection glare).

Quality

A refresh rate of at least 50 Hz for phosphor P4 (negative image), 60 Hz for P31 (negative image) or 70 Hz for positive image will minimize perceptible screen flicker. Other forms of display instability such as 'jitter' and 'swim' should also be minimised.

Characters

The preferred height of characters lies between 0·007 and 0·01 reading distance. Hence, 4 mm characters are suitable at 500 mm (Schmidtke 1980). The proportions of characters should follow the recommendations given above for general purpose alphanumerics. Character spacings of 10% symbol height are acceptable in high-quality displays but 25% may be required in other circumstances.

The legibility of dot matrix characters increases with the number of dots per character and decreases with their separation. A 5 × 7 matrix is only acceptable for capitals and numerals; lower case requires 7 × 9 but 9 × 11 is better.

Cursor

A box cursor flashing at 3–5 Hz is recommended. Cursor control keys should be clearly marked by a cross-shaped configuration of keys, preferably separated from the alphanumerics on the right-hand side of the board.

Contrast

Contrast between characters and background should be between 8:1 and 10:1 with brightness and contrast separately adjustable.

Segmented displays (LED and LCD)

A seven-segment LED or LCD configuration produces a perfectly acceptable set of numerals. Legibility will, however, be less than for other modes of presentation and a relatively high error rate is to be anticipated. This may, to some extent, be offset by increasing the size. Care must also be taken concerning viewing angles. If the quantity monitored is subject to transients, electronic averaging over at least 1 s is desirable. (No useful information is lost since the human observer cannot process data faster than this.)

A 14-segment display is only marginally acceptable for alphanumerics.

Synthesised speech

Although still in its infancy, computer-generated speech will soon be giving the designer enhanced possibilities for 'auditory displays' which have hitherto been largely limited to warning bells and buzzers or the occasional 'bleeping'. Typical present-day devices have a vocabulary of around 200 words, pre-recorded and digitally encoded. The original speaker is usually recognisable and quite subtle qualities, such as sincerity or condescension, may be maintained. Intonation and inflections are, however, lost, giving the voice a flatness which can become annoying (Michaelis 1980).

17.4. Combinations of displays and controls

Principles of panel layout

The following principles, which were originally conceived with reference to the control consoles of complex industrial/military installations, are applicable in many areas of design. The first four principles were stated in their classical form by McCormick (1970).

(*a*) Importance principle. The most important items (in terms of safety, performance, etc.) should be in the most advantageous locations.

(*b*) Frequency of use principle. The most frequently used items should be in the most advantageous locations. What are the most advantageous locations? For displays consider what will be the centre of visual interest and locate this item straight ahead, approximately 15° below eye level (as described in Section 8.3). All other high-priority displays should be within a cone of vision of 30° centred on this reference direction. Lower-priority displays can be up to 45° from the principal line. Beyond this zone the operator will only notice bold flashing lights. Priority hand controls should be in the central regions of the reach envelope.

(*c*) Functional principle. Items concerned with similar functions should be grouped together. Controls should be mounted next to their associated displays, positioned so that the operator's hand does not obscure his view.

(*d*) Sequence of use principle. Items commonly operated in sequence should be grouped together; preferably allowing the operator to progress from left to right or from top to bottom in an unbroken chain of operations. Where this is not possible, backtracking should be minimised.

(*e*) Where the monitoring of many displays is required, aid the operator's visual search by contriving all indicators to be in the same alignment in the neutral or 'safe' condition.

(*f*) All labels should be in the same relationship to the items they describe—immediately above is best.

(*g*) Distribute workload appropriately between left and right hands.

(*h*) Where related displays and controls must be physically separate (e.g., on vertical and horizontal panels), their arrangement should be as similar as possible.

17.5. *Stereotypes and compatibility*

Compatibility, as defined by McCormick (1970), "refers to the spatial, movement or conceptual features of stimuli and of responses, individually or in combination, which are most consistent with human expectations". The concept may be illustrated with a familiar example. Consider the domestic cooker which has a hob with four burners arranged in a square thus:

	Left	Right
Back	BL	BR
Front	FL	FR

The four controls are usually arranged in a row, on a vertical panel, on the front of the cooker; so that the user does not have to reach over the flames. (Less sensible arrangements are not unknown.) In what order should the controls appear? There are 4!, or 24, possibilities, some of which are downright absurd, but at least four of which are reasonable enough to be used on marketed products:

I:	BL,	FL,	FR,	BR
II:	FL,	BL,	BR,	FR
III:	BL,	FL,	BR,	FR
IV:	FL,	BL,	FR,	BR

Both Chapanis and Lindenbaum (1950) and Ray and Ray (1979) have investigated the relative merits of these arrangements, and found the first (I) to be superior in terms of speed, errors and user preference. Of the four possibilities tested, it is, therefore, said to have the greatest conceptual/spatial compatibility. However, the studies also showed that staggering the burners, or placing them in a straight line, reduced the error rate to zero.

The term motion stereotype is applied to compatibility relationships concerned with the expected consequences of control actions. See Loveless (1962) and Murrel (1969) for reviews. We may summarise the most important relationship thus:

control movements of clockwise rotation, displacement forwards (i.e., away from the operator), upwards or to the right

result in similar movements of physical objects (in any combination); or similar movements of dial indicators or other display devices (in any combination and in any pair of planes) which are indicative of increase in the rate, intensity, volume, level, etc., or sequential progression of some function.

Some of these verge on the self-evident. The clockwise-for-increase/progression convention is deeply ingrained. The hands of a clock rotate in the same direction as the Sun appears to rotate about the Earth and the shadow of the gnomon moves round a (horizontal) sundial. It is held to be bad form to reverse this 'natural' relationship in dealing cards or passing the port.

Are these relationships inherent or are they learned? Murrel (1969) inclines to the former view, although he acknowledges an overlay of learning. Ballantine (1983) on the basis of extensive studies of children concludes definitely in favour of the latter.

There are, however, major exceptions to the general rules:

(*i*) In Britain toggle switches, etc., are pressed down for on and up for off. In the USA the opposite convention is equally well established. Given the former we should expect a downward-for-increase relation for vertically mounted sliders—and, hence, if the slider is horizontal we should surely expect it to move towards us for increase. A major subset of relationships must therefore be culture bound.

(*ii*) The clockwise-for-increase stereotype is violated in the very large class of devices in which a screwthread (right-handed) opens and closes a valve. Hence, taps turn anti-clockwise for an increased flow and motor-cycle throttles for increased power.

(*iii*) In order to open a mortice lock one must turn the key or handle clockwise for a left-hand jam and anti-clockwise for a right-hand jam. Users commonly describe this as turning the handle away from the jam; and some books consider this to be a stereotypical arrangement in its own right (i.e., Woodson 1981). Locks on horizontal jams (such as desks and car boots) generally operate clockwise for open and there seems to be an underlying expectancy that this will be the case (Martin and Webb 1980). Some locks and catches open whichever way you turn them (e.g., 'Yale' locks). My own observations suggest that, in approaching these, some users conform to a clockwise-for-open stereotype and some to an away-from-the-jam stereotype.

(*iv*) If an object (e.g., doorknob) is grasped from above with the right hand, or from below with the left, it is natural to turn it clockwise; from above with the left, or below with the right, turn anti-clockwise. This is a biomechanical relationship (angle–torque curve of supination/pronation—see Section 8.3). It will therefore tend to over-ride other considerations.

(*v*) In novel and unpredictable situations users sometimes revert to a clockwise-for-anything stereotype—'if in doubt turn it clockwise'.

I should find it difficult to gainsay anyone who argued that stereotypes in the general sense do not exist. Not only are they learned rather than inherent but they are learned separately for screws, taps, keys, etc.—each class of device seems to generate its own stereotype.

17.6. *Conceptual compatibility*

I made reference at the beginning of this chapter to the notorious three-pointer altimeter—an illegible instrument which causes pilots to misjudge their altitude with disastrous results. Consider now another disaster—the nuclear power station failure at Three Mile Island, which was also due, in some measure, to ergonomic problems. In this case the display interface was readable but not intelligible. The operators drew the wrong inferences from the dial readings and their attempts to deal with the mounting crisis were inappropriate; due basically to a lack of comprehension of the system. I am reminded of Confucius:

> "If language is not correct, then what is said is not what is meant; if what is said is not what is meant, then what ought to be done remains undone."

At Three Mile Island the designers of the system failed to communicate with the operators in the language of the displays. (The problem was exacerbated by between 100 and 200 separate alarms going off all at once and adding to the general state of anarchy.)

In many contemporary systems (not only large-scale ones) the most pressing ergonomic problems are of this kind. Computers and word-processors provide innumerable examples. How may we supply the operator with an appropriate mental model of the system? I should like to be able to give operational answers to these questions and write simple guidelines for the designer. I am unable to do so since, to a great extent, the ergonomic study of these issues is not yet sufficiently advanced to permit rule-based rather than *ad hoc* solutions.

The Procrustean beds of the foreseeable future will be conceptual ones.

Appendix I
Glossary of terms

Abduction	A movement which generally takes place in the *coronal plane* (q.v.) causing a body member to move away from the mid-line (median plane). The opposite movement (which reverses abduction) is known as *adduction*.
Acromion	A bony prominence on the scapula (shoulder blade) which forms the palpable 'tip' of the shoulder.
Adduction	Opposite movement to *abduction* (q.v.).
Anterior	Towards the front of the body.
Anthropometry/ anthropometrics	The collection of numerical data concerning the physical characteristics of human beings, particularly size, shape and strength; the application of such data to the problems of design
Anthropometrist	A harmless drudge.
Cognitive ergonomics	The branch of *ergonomics* (q.v.) which deals with information processing, reasoning, problem-solving activities, etc.
Coronal plane	Any vertical plane perpendicular to the *median plane* of the body (q.v.).
Distal	Away from the trunk (concerning limb segments).
Ergonomics	The application of scientific information concerning human beings (and scientific methods for obtaining such information) to the problems of design; user-centred design. Known in the USA as *human factors*.
Extension	Opposite movement to *flexion* (q.v.).
Fitting trial	A type of psychophysical experiment. The dimensions of an object (for example a workstation) are systematically varied and the subjective impressions of users are recorded.
Flexion	A movement which takes place in the *sagittal plane* (q.v.), causing the bending of the limbs or the opposition of morphologically ventral surfaces (i.e., the face, the front of the chest, abdomen and thigh, the back of knee and lower leg, the sole

of the foot, the palm of the hand and the surface of the forearm and upper arm which is continuous with it). The foetus lies in a flexed position within the womb. The opposite movement to flexion (i.e., straightening the limbs, etc.) is known as *extension*.

Functionalism — School, style or philosophy of design which holds that the form of an object should be determined by its function. If the analysis of function is limited to superficial characteristics, such as cost and construction, an object may result which appears 'functional' (by virtue of the lack of decoration or sensitivity) but is no more than marginally fit for its purpose. Ergonomics aims to remedy these defects by a vigorous consideration of the capacities, needs and tastes of the human user.

Human factors — Synonym for *ergonomics* (q.v.), main in the USA.

Interface — Imaginary surface across which information is transmitted from operator to machine (by controls) and vice versa (by displays).

Kyphosis — A curve of the spine which is convex to the rear.

Lateral — Away from the *median plane* (q.v.).

Lordosis — A curve of the spine which is concave to the rear.

Medial — Towards the *median plane* (q.v.).

Median plane — The plane which divides the body into equal right and left halves (also known as the mid-sagittal plane). Also the plane which runs down the midline of a limb—as from the shoulder to the tip of the middle finger.

Metacarpals — The five bones of the palm of the hand (the metacarpal of the thumb is movable, the remainder are almost fixed).

Method of limits — Originally an experimental technique in psychophysics. The term is sometimes applied (by analogy) to the general method of anthropometrics in which an 'imaginary' *fitting trial* is conducted using data as a 'substitute' for subjects.

Popliteal — Pertaining to the popliteal fossa which is the hollow at the back of the knee where the underside of the thigh meets the back of the lower leg in the sitting position.

Posterior — Towards the back of the body.

Pronation — Opposite movement to *supination* (q.v.).

Proximal — Towards the trunk (concerning limb segments).

Psychophysics — The measurement of sensation.

Sagittal plane — Any plane parallel to the *median plane* of the body (q.v.).

SGO — Stunning glimpse of the obvious: commonly encountered in handbooks of ergonomics.

SRP — Seat reference point: defined by the intersection of (*i*) the median plane of the body, (*ii*) the backrest (or in its absence a vertical plane tangent to both shoulder blades) and (*iii*) the (compressed) seat surface.

Stereotype — The expected mode of operation of a device; particularly the expected consequences of a particular control movement.

Supination — A rotatary movement of the forearm causing the palm of the hand to face upwards (equivalent to a clockwise turn with the right hand or an anti-clockwise turn with the left). The opposite movement is called *pronation*.

Systems ergonomics — The branch of *ergonomics* (q.v.) which deals with the human operator as a component of a *man–machine system* or a *socio-technical system* (q.v.).

Torque — The turning effect of a force—numerically equal to the product of the magnitude of force and the perpendicular distance between the line of action of the force and the axis of rotation.

Ulnar — Pertaining to the ulna (bone); on the little finger side of the forearm.

Appendix II
Units and conversion factors

Since 1973 the SI system of units has been officially preferred by the scientific world. Its use is outlines in various standards (e.g., BS 5775, BS 5555, ISO 1000, ISO 31/0). In spite of this, its acceptance has been by no means universal. Physiologists tend to prefer the kilogram force or kilopond to the Newton and in the USA large segments of the scientific community have ignored SI units entirely and continued to employ Imperial inches and pounds. No reasonable person would use the radian as an angular measure in preference to the degree (except as an aid to calculation). The present author has adopted a middle path by using millimetres, kilogram force and degrees. (Body weight is quoted in kilograms, since that is the way that all metric weighing machines are calibrated. The purist may call it 'body mass' or convert it to Newtons.)

For the benefit of those readers who prefer different units, the following conversion factors are provided:

Length

1 inch (in) = 25·4 millimetres (mm)
10 mm = 0·394 in
1 metre (m) = 39·4 in

Mass

1 kilogram (kg) = 2·2 pounds (lb)
= 35·2 ounces (oz)
1 pound (lb) = 0·454 kg

Force/weight

1 kilogram force (kgf) = 1 kilopond (kp)
= 9·81 newtons (N)
= 2·2 pounds force (lbf)
1 lbf = 4·45 N

Torque

1 kilogram force metre (kgf m) = 9·81 Newton metres (N m)
= 86·7 inch pounds
1 inch pound = 0·0115 kgf m
= 0·113 N m

Angle

1 radian = $360/2\pi$ degrees (°)
= 57·3 degrees (°)

References

Abraham, S. (1979), Weight and height of adults 18–74 years of age, United States, 1971–1974. *Vital and Health Statistics, Series 11*, No. 211, US Department of Health Education and Welfare, MD.

Adams, G. A. (1961), A comparative anthropometric study of hard labor during youth as a stimulator of physical growth of young colored women. *Research Quarterly AAHPER*, **9**, 102–108.

Andersson, G. B. J. (1979), Low back pain in industry: epidemiological aspects. *Scandinavian Journal of Rehabilitation Medicine*, **11**, 163–168.

Andersson, G. B. J., Ortengren, R., Nachemson, A. and Elfstrom, G. (1974), Lumbar disc pressure and myoelectric back muscle activity during sitting. I, Studies on an experimental chair. *Scandinavian Journal of Rehabilitation Medicine*, **3**, 104–114.

Andersson, M., Hwang, S. G. and Green, W. T. (1965), Growth of the normal trunk in boys and girls during the second decade of life. *Journal of Bone and Joint Surgery*, **47A**, 1554–1564.

Andrew, I. and Manoy, R. (1972), Anthropometric survey of British Rail footplate staff. *Applied Ergonomics*, **3**, 132–135.

Archer, J. and Lloyd, B. (1982), *Sex and Gender* (Harmondsworth: Penguin).

Ashley-Montagu, M. F. (1960), *An Introduction to Physical Anthropology*, 3rd Edition (Springfield, IL: Charles C. Thomas).

Ayoub, M. M. and McDaniel, J. W. (1973), Effects of operator stance on pushing and pulling tasks. *AIIE Transactions*, **6**, 185–195.

Babbs, F. W. (1979), A design layout method for relating seating to the occupant and vehicle. *Ergonomics*, **22**, 227–234.

Backwin, H. and McLaughlin, S. D. (1964), Increase in stature—is the end in sight? *Lancet*, **ii**, 1195–1197.

Ballantine, M. (1983), Well, how *do* children learn population sterotypes? *Proceedings of the Ergonomics Society's Conference 1983*, Ed. K. Coombes, p. 1 (London: Taylor & Francis).

Barkla, D. (1961), The estimation of body measurements of the British population in relation to seat design. *Ergonomics*, **4**, 123–132.

Barnes, R. M. (1958), *Motion and Time Study* (New York: Wiley).

Barter, T., Emmanuel, I. and Truett, B. (1957), A statistical evaluation of joint range data. WADC Technical Note 53–311, Wright Patterson Airforce Base, OH.

Batogowska, A. and Slowikowski, J. (1974), Anthropometric atlas of the Polish adult population for designer use. Instytut Wzornictwa Przemystowego, Warsaw (in Polish).

Bauer, D. and Cavonius, C. R. (1980), Improving the legibility of visual display units through contrast reversal. In Grandjean and Vigliani (1980), op.cit., pp. 137–142.

Bendix, T. and Hagberg, M. (1984), Trunk posture and load on the trapezius muscle whilst sitting at sloping desks. *Ergonomics*, **27**, 873–882.

Bennett, C. (1977), *Spaces for People—Human Factors in Design* (Englewood Cliffs: Prentice Hall).

Bittner, A. C., Dannhaus, D. M. and Roth, J. T. (1975), Workplace—accommodated percentage evaluation: model and preliminary results. In *Improved Seat, Console and Workplace Design*, ed. M. M. Ayoub and C. G. Halcomb, Pacific Missile Test Center, Point Mugu, CA 93042.

Boas, F. (1912), *Changes in Bodily Form of Descendents of Immigrants* (New York: Columbia University Press).

Borkan, G. A., Hults, D. E. and Glynn, R. J. (1983), Role of longitudinal change and secular trend in age differences in male body dimensions. *Human Biology*, **55**, 629–641.

Borkan, G. A. and Norris, A. H. (1977), Fat redistribution and the changing body dimensions of the adult male. *Human Biology*, **49**, 495–514.

Branton, P. (1984), Backshapes of seated persons—how close can the interface be designed? *Applied Ergonomics*, **15**, 105–107.

Brown, C. H. and Wilmore, J. H. (1974), The effects of maximal resistance training on the strength and body composition of women athletes. *Medicine and Science in Sports*, **6**, 174–177.

Brown, C. R. and Schaum, D. L. (1980), User-adjusted VDU parameters. In Grandjean and Vigliani (1980), op.cit., pp. 195–200.

Brunswic, M. (1981), How seat design and task affect the posture of the spine in unsupported sitting. MSc. Dissertation, Ergonomics Unit, UCL, 26 Bedford Way, London WC1.

BS 3042: 1971 (1980), *Standard Test Fingers and Probes for Checking Protection against Electrical, Mechanical and Thermal Hazards* (London: British Standards Institution)

BS 3693: Part 1: 1964, *Instruments for Bold Presentation and Rapid Reading* (London: British Standards Institution).

BS 3705: 1972, *Recommendations for Provision of Space for Domestic Kitchen Equipment* (London: British Standards Institution).

BS 5304: 1975, *Code of Practice for Safeguarding of Machinery* (London: British Standards Institution).

BS 5555: 1981, *Specification for SI Units and Recommendations for the Use of their Multiples and Certain Other Units* (= ISO 1000) (London: British Standards Institution).

BS 5619: 1978, *Code of Practice for Design of Housing for the Convenience of Disabled People* (London: British Standards Institution).

BS 5775: Part 0: 1982, *Specification for Quantities, Units and Symbols (= ISO 31/0)* (London: British Standards Institution).

BS 5940: Part 1: 1980, *Office Furniture—Specification for Design and Dimensions of Office Workstations, Desks, Tables and Chairs* (London: British Standards Institution).

BS 5959: 1980, *Specification for Key Numbering System and Layout Charts for Keyboards on Office Machines (= ISO 4169)* (London: British Standards Institution).

BS 6222: Part 1: 1982, *Domestic Kitchen Equipment. Specification for Co-ordinating Dimensions* (London: British Standards Institution).

BS AU176 (1980), *Method for Establishment of Eyellipses for Driver's Eye Location* (= ISO 4513) (London: British Standards Institution).

Caillet, R. (1977), *Soft Tissue Pain and Disability* (Philadelphia: F. A. Davis).

Cakir, A., Hart, D. J. and Stewart, T. F. M. (1980), *Visual Display Terminals* (Chichester: Wiley).

Cameron, I. (1984), What is common sense? Unpublished paper presented to the London Ergonomics Group.

Cameron, N. (1979), The growth of London schoolchildren 1904–1966: an analysis of secular trend and intra-county variation. *Annals of Human Biology*, **6**, 505–525.

Cameron, N., Tanner, J. M. and Whitehouse, R. A. (1982), A longitudinal analysis of the growth of limb segments in adolescence. *Annals of Human Biology*, **9**, 211–220.

Chaffin, D. B. and Andersson, G. B. J. (1984), *Occupational Biomechanics* (New York: Wiley).

Chapanis, A. (1959), *Research Techniques in Human Engineering* (Baltimore: The Johns Hopkins Press).

Chapanis, A. and Lindenbaum, L. (1950), A reaction time study of four central display linkages. *Human Factors*, **1**, 1–17.

Clark, T. S. and Corlett, E. N. (1984), *The Ergonomics of Workspaces and Machines—A Design Manual* (London: Taylor & Francis).

Conrad, R. and Hull, A. J. (1968), The preferred layout for numeral data entry keysets. *Ergonomics*, **11**, 165–174.

Corlett, E. N. (1983), Analysis and evaluation of working posture. In *Ergonomics of Workstation Design*, ed. T. O. Kvalseth (London: Butterworths), Chapter 1.

Corlett, E. N. and Manenica, I. (1980), The effects and measurement of working postures. *Applied Ergonomics*, **11**, 7–16.

Cox, C. F. (1984), An investigation of the dynamic anthropometry of the seated workplace. MSc. Dissertation, University College London, 26 Bedford Way, London WC1.

Cyriax, J. H. (1978), *Textbook of Orthopaedic Medicine*, 7th edition (London: Ballière Tindall).

Dalassio, D. J. (ed.) (1980), *Wolff's Headache and Other Head Pain* (Oxford: Oxford University Press).

Damon, A. (1973), Ongoing human evolution. In *Heredity and Society*. eds. I. H. Porter and R. E. Skalko (New York: Academic Press), pp. 45–74.

Damon, A., Seltzer, C. C., Stoudt, H. W. and Bell, B. (1972), Age and physique in healthy white veterans at Boston. *Journal of Gerontology*, **27**, 202–208.

Damon, A., Stoudt, H. W. and McFarland, R. A. (1966), *The Human Body in Equipment Design* (Cambridge, MA: Harvard University Press).

Das, B. and Grady, R. M. (1983), The normal working area in the horizontal plane. A comparative analysis between Farley's and Squires' concepts. *Ergonomics*, **26**, 449–459.

Davies, B. T., Abada, A., Benson, K., Courtney, A. and Minto, I. (1980 a), Female hand dimensions and guarding of machines. *Ergonomics*, **23**, 79–84.

Davies, B. T., Abada, A., Benson, K., Courtney, A. and Minto, I. (1980 b), A comparison of hand anthropometry of females in three ethnic groups. *Ergonomics*, **23**, 179–182.

Davis, P. R. and Stubbs, D. A. (1977), Safe levels of manual forces for young males. *Applied Ergonomics*, **8**, 141–150, 219–228.

Davis, P. R. and Stubbs, D. A. (1978), Safe levels of manual forces for young males. *Applied Ergonomics*, **9**, 33–37.

Dempster, W. T. (1955), Space requirements of the seated operator: geometrical, kinematic and mechanical aspects of the body with special reference to the limbs. WADC Tech. Note 55–159, Wright Patterson Air Force Base, OH.

Department of Defense (1981), Human Engineering Design Criteria for Military Systems, Equipment and Facilities. MIL-STD-1472C (Washington, DC: Dept of Defense).

Department of Education and Science (1972), *British School Population Dimensional Survey*, Building Bulletin 46, Department of Education and Science (London: HMSO).

Department of Education and Science (1985), *Body Dimensions of the School Population*. Building Bulletin 62, Department of Education and Science (London: HMSO).

Department of the Environment (1972) *Space in the Home*. Design Bulletin 6, Department of the Environment (London: HMSO).

Diffrient, N., Tilley, A. and Bardagjy, J. C. (1974), *Humanscale*, 1/2/3 (Cambridge, MA: MIT Press).

Diffrient, N., Tilley, A. and Harman, D. (1981), *Humanscale* 4/5/6/7/8/9 (Cambridge, MA: MIT Press).

DIN 31001 Part 1 (1983), *Safety Distances for Adults and Children* (Berlin: Deutsches Institut für Normung).

Dreyfuss, H. (1971), *The Measure of Man* (New York: Whitney).

Drillis, R. and Contini, R. (1966), Body segment parameters, Technical Report No. 1166.03, School of Engineering and Science, New York University.

Drury, C. G. (1980), Handles for manual materials handling. *Applied Ergonomics*, **11**, 35–42.

Drury, C. G. and Francher, M. (1985), Evaluation of a forward-sloping chair. *Applied Ergonomics*, **16**, 41–47.

Ducharme, R. E. (1975), Problem tools for women. *Industrial Engineering*, **7**, 46–50.

Ducharme, R. E. (1977), Women workers rate male tools inadequate. *Human Factors Society Bulletin*, **20**, 1–2.

Duncan, J. and Ferguson, D. (1974), Keyboard operating posture and symptoms in operating. *Ergonomics*, **17**, 651–662.

Durnin, J. V. G. A. and Rahaman, M. M. (1967), The assessment of the amount of fat in the human body from measurements of skinfold thickness. *British Journal of Nutrition*, **21**, 681–689.

Durnin, J. V. G. A. and Womersley, J. (1974), Body fat assessed from total body density and its estimation from skinfold thickness: measurements on 481 men and women aged from 16 to 72 years. *British Journal of Nutrition*, **32**, 77–97.

Elmfeldt, G., Wise, C., Beresten, H. and Olsson, A. (1981), *Adapting Work Sites for People with Disabilities.* The Swedish Institute for the Handicapped, S–16126 Bromma, Sweden/World Rehabilitation Fund, 400 East 34th Street, New York.

Ericsson (1983), *Ergonomic Principles in Office Automation* (Bromma: Ericsson Information Systems AB).

Eveleth, P. B. (1975), Differences between ethnic groups in sex dimorphism of adult height. *Annals of Human Biology*, **2**, 35–39.

Eveleth, P. B. and Tanner, J. M. (1976), *Worldwide Variation in Human Growth* (Cambridge University Press).

Floyd, W. F., Guttman, L., Wycliffe-Noble, C., Parkes, M. A. and Ward, B. A. (1966), A study of the space requirements of wheelchair users. *Paraplegia*, May 1966, 24–37.

Floyd, W. F. and Ward, J. S. (1966), Posture in industry. *International Journal of Production Research*, **5**, 213–224.

Friedlander, J. S., Costa, P. T., Bosse, R., Ellis, E., Rhoads, J. G. and Stoudt, H. W. (1977), Longitudinal physique changes among healthy white veterans at Boston. *Human Biology*, **49**, 541–558.

Fruin, J. (1971), *Pedestrian Planning and Design* (New York: Metropolitan Association of Urban Designers and Environmental Planners).

Galer, D. M., Baines, A. and Simmonds, G. (1980), Ergonomic aspects of electronic dashboard instrumentation. In Oborne and Levis (1980), op.cit., Vol. 1, pp. 301–309.

Garner, D. M., Garfinkle, P. E., Schwartz, D. and Thompson, M. (1980), Cultural expectations of thinness in women. *Psychological Reports*, **47**, 483–491.

Garret, J. W. (1971), The adult human hand: some anthropometric and biomechanical considerations. *Human Factors*, **13**, 117–131.

Goldsmith, S. (1976), *Designing for the Disabled*, 3rd edition (London: RIBA).

Goldstein, H. (1971), Factors influencing the height of 7 year old children. *Human Biology*, **43**, 92–111.

Gooderson, C. Y. and Beebee, M. (1976), Anthropometry of 500 infantrymen 1973–1974. Report APRE 17/76, Army Personnel Research Establishment, Farnborough, Hants.

Gooderson, C. Y. and Beebee, M. (1977), A comparison of the anthropometry of 100 guardsmen with that of 500 infantrymen, 500 RAC servicemen and 2000 RAF aircrew. Report APRE 37/76, Army Personnel Research Establishment, Farnborough, Hants.

Gooderson, C. Y., Knowles, D. J. and Gooderson, P. M. E. (1982), The hand anthropometry of male and female military personnel. APRE Memorandum 82M510, Army Personnel Research Establishment, Farnborough, Hants.

Gould, S. J. (1984), *The Mismeasure of Man* (Harmondsworth: Penguin).

Grahame, R. and Jenkins, J. M. (1972), Joint hypermobility—asset or liability? A study of joint mobility in ballet dancers. *Annals of the Rheumatic Diseases*, **31**, 109–111.

Grandjean, E. (1973), *Ergonomics of the Home* (London: Taylor & Francis).

Grandjean, E. (1981), *Fitting the Task to the Man: An Ergonomic Approach*, 2nd edition (London: Taylor & Francis).

Grandjean, E. and Hünting, W. (1977), Ergonomics of posture—review of various problems of standing and sitting posture. *Applied Ergonomics*, **8**, 135–140).

Grandjean, E., Hünting, W., Maeda, K. and Läubli, Th. (1983), Constrained postures at office workstations. In *Ergonomics of Workstation Design*, ed. T. O. Kvalseth (London: Butterworths), Chapter 2.

Grandjean, E., Hünting, W. and Nishiyama, K. (1984), Preferred VDT workstation settings, body posture and physical impairments. *Applied Ergonomics*, **15**, 99–104.

Grandjean, E., Nishiyama, K., Hünting, W. and Piderman, M. (1984). A laboratory study on preferred and imposed settings of a VDT workstation. *Behaviour and Information Technology*, **3**, 289–304.

Grandjean, E. and Vigliani, E. (eds) (1980), *Ergonomic Aspects of Visual Display Terminals* (London: Taylor & Francis).

Greulich, H. (1957), A comparison of the physical growth and development of American born and native Japanese children. *American Journal of Physical Anthropology*, **15**, 489–516.

Grieve, D. W. (1984), The influence of posture on power output generated in single pulling movements. *Applied Ergonomics*, **15**, 115–116.

Grieve, D. W. and Pheasant, S. T. (1982), Biomechanics. In *The Body at Work*, ed. W. T. Singleton (Cambridge University Press), Chapter III.

Griew, S. (1969), *Adaptation of Jobs for the Disabled* (Geneva: International Labour Office).

Haigh, R. (1984), An ergonomic assessment of British Standard 5810: 1979. Access for the disabled to buildings, through a survey of architects. Project Report, Department of Human Sciences, Loughborough University of Technology.

Hall, E. T. (1969), *The Hidden Dimension* (New York: Doubleday).

Hamer, M. (1985), How speed kills on our roads. *New Scientist*, 21 February 1985, 10–11.

Hamilton, N. (1983), Optimal location of ovens for wheelchair users: an ergonomic approach. MSc. Dissertation, Ergonomics Unit, University College London, 26 Bedford Way, London WC1.

Hammond, D. C. and Roe, R. W. (1972), SAE controls reach study, 73061, Society of Automotive Engineers.

Hansen, R. and Cornog, D. Y. (1958), Annotated bibliography of applied physical anthropology in human engineering. WADC Technical Report 56–30, Wright Patterson Airforce Base, OH.

Harris, A. I. (1971), *Handicapped and Impairerd in Great Britain: Part I* (London: HMSO).

Hartley, J. and Burnhill, P. (1977), Fifty guidelines for improving instructional text. *Programmed Learning and Educational Technology*, **14**, 65–73.

Haslegrave, C. M. (1979), An anthropometric survey of British drivers. *Ergonomics*, **22**, 145–154.

Haslegrave, C. M. and Hardy, R. N. (1979), *Anthropometric Profile of the British Car Occupant*. Motor Industry Research Association.

Haslegrave, C. M. and Searle, J. A. (1980), Anthropometric considerations in seat belt design and testing. In Oborne and Levis (1980), op. cit., Vol. 1, pp. 374–382.

Healy, M. J. R. (1962), The effect of age-grouping on the distribution of a measurement effected by growth. *American Journal of Physical Anthropology*, **20**, 49–50.

Health and Safety Commission (1982), *Proposals for Health and Safety (Manual Handling of Loads). Regulations and Guidance.* Health and Safety Commission Consultative Document (London: HMSO).

Health and Safety Executive (1975–1980 a), *Health and Safety—Manufacturing and Service Industries* (London: HMSO).

Health and Safety Executive (1978–1980 b), *Health and Safety Statistics* (London: HMSO).

Health and Safety Executive (1983), *Visual Display Units* (London: HMSO).

Hertzberg, H. T. E. (1968), the conference on standardization of anthropometric techniques and terminology. *American Journal of Physical Anthropology*, **28**, 1–16.

Hertzberg, H. T. E., Churchill, E., Dupertuis, C. W., White, R. M. and Damon, A. (1963), *Anthropometric Survey of Greece, Turkey and Italy* (Oxford: Pergamon).

Hettinger, T. (1961), *Physiology of Strength* (Springfield, IL: Charles C. Thomas).

HFDNE (1971), *Human Factors for Designers of Naval Equipment* (London: Admiralty).

Hira, D. S. (1980), An ergonomic appraisal of educational desks. *Ergonomics*, **23**, 213–221.

HM Chief Inspector of Factories (1924–1974), *Annual Report* (London: HMSO).

Hogarth, W. (1753), *The Analysis of Beauty, Written with a View to Fixing the Fluctuating Ideas of Taste* (London).

Hornibrook, F. A. (1934). *The Culture of the Abdomen* (New York: Doubleday).

Hult, L. (1954), Cervical, dorsal and lumbar spinal syndromes. *Acta Orthopaedica Scandinavia*, Supplement 17.

Hünting, W., Grandjean, E. and Maeda, K. (1980), Constrained postures in accounting machine operators. *Applied Ergonomics*, **14**, 145–149.

Hünting, W., Läubli, Th. and Grandjean, E. (1981), Postural and visual loads at VDT workplaces. 1, Constrained postures. *Ergonomics*, **24**, 917–931.

Ikai, M. and Fukanaga, T. (1968), Calculation of muscle strength per unit cross-sectioned area of human muscle by means of ultrasonic measurement. *Int. Z. Angew. Physiol. Einschl. Arbeits-Physiol.*, **26**, 26–32.

Inglemark, B. E. and Lewin, T. (1968), Anthropometrical studies on Swedish women. *Acta Morphologica Neerlando-Scandinavica*, **III(2)**, 145–166.

Institute for Consumer Ergonomics (1983), Seating for elderly and disabled people. Report No. 2, Anthropometric Survey, University of Technology, Loughborough.

ISO 31/0 (1981), *General Principles Concerning Quantities, Units and Symbols* (= BSS 5775) (Geneva: International Standards Organization).

ISO 1000 (1981), *SI Units and Recommendations for the Use of Their Multiples and of Certain Other Units* (= BS 5555) (Geneva: International Standards Organization).

ISO 3958 (1977), *Road Vehicles—Passenger Cars—Driver Hand Control Reach* (Geneva: International Standards Organization).

ISO 4040 (1983), *Road Vehicles—Passenger Cars—Location of Hand Controls, Indicators and Tell-tales* (Geneva: International Standards Organization).

ISO 4169 (1979), *Office Machines—Keyboards—Key Numbering System and Layout Charts* (= BS 5959) (Geneva: International Standards Organization).

ISO 4513 (1978), *Road Vehicles—Visibility—Method for Establishment of Eyellipses for Drivers' Eye Location* (= BS AU176) (Geneva: International Standards Organization).

ISO 6549 (1980), *Road Vehicles—Procedure for H-point determination* (Geneva: International Standards Organization).

James, J. (1951), A preliminary study of the size determinant in small group interaction. *American Sociological Review*, **16**, 474–477.

Jones, J. C. (1963), Fitting trials—a method of fitting equipment dimensions to variation in the activities, comfort requirements and body sizes of users. *The Architects Journal*, 6 February 1963, 321–325.

Jones, J. C. (1969), Methods and results of seating research. *Ergonomics*, **12**, 171–181.

Kapanji, I. A. (1974), *The Physiology of the Joints* (Edinburgh: Churchill Livingstone).

Kaplan, B. A. (1954), Environment and human plasticity. *American Anthropology*, **56**, 780–800.

Karagelis, I. (1982), An ergonomic evaluation of proposed kitchen worktops for wheel-chair disabled. MSc Dissertation, Ergonomics Unit, University College London, 26 Bedford Way, London WC1.

Kember, P., Ainsworth, L. and Brightman, P. (1981), *A Hand Anthropometric Survey of British Workers*. Ergonomics Laboratory, Cranfield Institute of Technology.

Kennedy, K. W. (1964), Reach capability of the USAF population. Phase I, the outer boundaries of grasping reach envelopes for the shirt sleeved, seated operator. Report AMRL-TDR-64-59, Wright Patterson Airforce Base, OH.

Kira, A. (1976), *The Bathroom* (Harmondsworth: Penguin).

Klafs, C. E. and Lyon, M. J. (1978), *The Female Athlete—a Coaches Guide to Conditioning and Training* (St. Louis: C. V. Mosby).

Knight, I. (1984), *The Heights and Weights of Adults in Great Britain* (London: HMSO).

Koblianski, E. and Arensburg, B. (1977), Changes in morphology of human populations due to migration and selection. *Annals of Human Biology*, **4**, 57–71.

Kroemer, K. H. E. (1971), Foot operation of controls. *Ergonomics*, **14**, 333–361.

Kroemer, K. H. E. (1972), Human engineering—the keyboard. *Human Factors*, **14**, 51–63.

Kukkonen, R., Luopajarvi, T. and Riihimaki, V. (1983), Prevention of fatigue amongst data entry operators. In *Ergonomics in Workstation Design*, ed. T. O. Kvalseth (London: Butterworths), Chapter 3.

Lawson, B. (1980), *How Designers Think* (London: Architectural Press).

Le Carpentier, E. F. (1969), Easy chair dimensions for comfort. *Ergonomics*, **12**, 328–337.

Le Corbusier (1961), *The Modulor—A Harmonious Measure to the Human Scale, Universally Applicable to Architecture and Mechanics* (London: Faber and Faber).

Lehman, G. (1958), Physiological basis of tractor design. *Ergonomics*, **1**, 197–205.
Lewin, T. (1969), Anthropometric studies on Swedish industrial workers when standing and sitting. *Ergonomics*, **12**, 883–902.
Life, M. A. and Pheasant, S. T. (1984), An integrated approach to the study of posture in keyboard operation. *Applied Ergonomics*, **15**, 83–90.
Lindgren, G. (1976), Height, weight and menarche in Swedish schoolchildren in relation to socio-economic and regional factors. *Annals of Human Biology*, **3**, 510–528.
Little, K. B. (1965), Personal space. *Journal of Experimental Social Psychology*, **1**, 237–247.
Loveless, N. E. (1962), Directional motion stereotypes—a review. *Ergonomics*, **5**, 357–381.
McClelland, I. L. and Ward, J. S. (1976), Ergonomics in relation to sanitary ware design. *Ergonomics*, **19**, 465–478.
McClelland, I. L. and Ward, J. S. (1982), The ergonomics of toilet seats. *Human Factors*, **24**, 713–725.
McCormick, E. S. (1970), *Human Factors Engineering* (New York: McGraw-Hill).
McWhirter, N. D. (1984), *Guinness Book of Records*, 30th edition (Enfield: Guinness Superlatives).
Maeda, K. (1977), Occupational cervicobrachial disorder and its causative factors. *Journal of Human Ergology*, **6**, 193–202.
Malina, R. M. and Zavaleta, A. N. (1976), Androgyny of physique in female track and field athletes. *Annals of Human Biology*, **3**, 441–446.
Mandal, A. C. (1976), Work chair with tilted seat. *Ergonomics*, **19**, 157–164.
Mandal, A. C. (1981), The seated man (Homo sedens). The seated work position, theory and practice. *Applied Ergonomics*, **12**, 19–26.
Marquer, P. and Chalma, M. C. (1961), L'evolution des characteres morphologiques en function de l'age chez 2089 francaises de 20 a 91 ans. *Bulletin et Memoires de la Societe d'Anthropologie de Paris*, **XI(2)**, 1–78.
Martin, W. E. (1960), Children's body measurements for planning and equipping schools. Special Publication No. 4, US Department of Health, Education and Welfare, MD.
Martin, J. and Webb, R. D. G. (1980), Response stereotypes for key turning. *Proceedings of the Annual Conference of the Human Factors Association of Canada*, pp. 105–107.
Medawar, P. B. (1944), Size, shape and age. In *Essays on Growth and Form Presented to D'Arcy Wentworth Thompson* (Oxford: Clarendon Press), pp. 155–187.
Melzack, R. and Wall, P. (1982), *The Challenge of Pain* (Harmondsworth: Penguin).
Meredith, H. W. (1976), Findings from Asia, Australia, Europe and North America on secular change in mean height of children, youths and young adults. *American Journal of Physical Anthropology*, **44**, 315–326.
Miall, W. E., Ashcroft, M. T., Lovell, H. G. and Moore, F. (1967), A longitudinal study of the decline of adult height with age in two Welsh communities. *Human Biology*, **39**, 445–454.
Michaelis, P. R. (1980), An ergonomist's introduction to synthesized speech. In Oborne and Levis (1980), op. cit., Vol. 1, pp. 291–294.
MIIS (1984), Motor industry information service fact sheets. In *The Automania Fact File* (Birmingham: Central Television).
Miller, C. D. (1961), Stature and build of Hawaii-born youth of Japanese ancestry. *American Journal of Physical Anthropology*, **19**, 159–171.
Miller, D. I. and Nelson, R. C. (1973), *Biomechanics of Sport* (Philadelphia: Lea and Febiger).
Ministry of Defence, London (1984), Defence Standard. Human Factors for Designers of Equipment. Part 3: Body Strength and Stamina. MoD 00-25, Part 3 (London: Ministry of Defence).
Mitchie, D. and Johnston, R. (1984), *The Creative Computer—Machine Intelligence and Human Knowledge* (Harmondsworth: Penguin).
Montegriffo, V. M. E. (1968), Height and weight of a United Kingdom adult population with a review of anthropometric literature. *Annals of Human Genetics*, **31**, 389–398.
Montoye, H. J. and Lamphier, D. E. (1977), Grip and arm strengths in males and females. *Research Quarterly of the American Association for Health, Physical Education and Recreation*, **48**, 109–120.
Morris, J. N., Heady, J. A. and Raffle, P. A. B. (1956), Physique of London busmen—epidemiology of uniforms. *Lancet*, 15 September, 569–570.

Murrel, K. F. H. (1969), *Ergonomics—Man and His Working Environment* (London: Chapman and Hall).

Napier, J. R. (1956), The prehensile movements of the human hand. *Journal of Bone and Joint Surgery*, **38B,** 902–913.

NASA (1978), *Anthropometric Source Book.* NASA Defence Publication No. 1024 (US National Aeronautics and Space Administration).

NIOSH (1981), *Work Practices Guide for Manual Lifting.* National Institute for Occupational Safety and Health, Cincinnati, OH 45226.

Noble, J. (1982), *Activity and Spaces, Dimensional Data for Housing Design* (London: The Architectural Press).

Oborne, D. J. (1981), *Ergonomics at Work* (Chichester: Wiley).

Oborne, D. J. and Heath, T. O. (1979), The role of social space requirements in ergonomics. *Applied Ergonomics*, **10,** 99–103.

Oborne, D. J. and Levis, J. A. (eds) (1980), *Human Factors in Transport Research* (London: Academic Press).

OPCS (1981), *Adult Heights and Weights Survey.* OPCS Monitor ref. 5581/1 (London: Office of Population Census and Surveys).

Panofsky, E. (1970), *Meaning in the Visual Arts* (Harmondsworth: Penguin).

Pepermans, R. G. and Corlett, E. N. (1983), Cross-modality matching as a subjective assessment technique. *Applied Ergonomics*, **14,** 169–176.

Pheasant, S. T. (1982 a), A technique for estimating anthropometric data from the parameters of the distribution of stature. *Ergonomics*, **25,** 981–992.

Pheasant, S. T. (1982 b), Anthropometric estimates for British civilian adults. *Ergonomics*, **25,** 993–1001.

Pheasant, S. T. (1983), Sex differences in strength—some observations on their variability. *Applied Ergonomics*, **14,** 205–211.

Pheasant, S. T. (1984 a), Human proportions—sex, age and ethnic differences. In *Contemporary Ergonomics 1984*, ed. E. D. Megaw (London: Taylor & Francis), pp. 142–147.

Pheasant, S. T. (1984 b), *Anthropometrics—an Introduction for Schools and Colleges.* BSI Educational Publication 7310 (London: British Standards Institution).

Pheasant, S. T. and Harris, C. M-T. (1982), Human strength in the operation of tractor pedals. *Ergonomics*, **25,** 53–63.

Pheasant, S. T. and O'Neill, D. (1975), Performance in gripping and turning. *Applied Ergonomics*, **6,** 205–208.

Pheasant, S. T. and Scriven, J. G. (1983), Sex differences in strength—some implications for the design of hand tools. In *Proceedings of the Ergonomics Society's Conference 1983*, ed. K. Coombes (London: Taylor & Francis), pp. 9–13.

Pheasant, S. T., Grieve, D. W., Rubin, T. and Thomson, S. J. (1982), Vector representations of human strength in whole body exertion. *Applied Ergonomics*, **13,** 139–144.

Plagenhof, S. (1971), *Patterns of Human Motion—A Cinematographic Analysis* (Englewood Cliffs, NJ: Prentice Hall).

Radl, G. W. (1980), Experimental investigations for optimal presentation mode and colour of symbols on the CRT-screen. In Grandjean and Vigliani (1980), op. cit., pp. 271–276.

Ramaprakash, D. (ed.) (1984), *Social Trends* (London: HMSO).

Ramazzini, B. (1713), *De Morbis Artificum.* Translated by W. C. Wright (1940), *Diseases of Workers* (University of Chicago Press).

Ray, R. D. and Ray, W. D. (1979), An analysis of domestic cooker control design. *Ergonomics*, **22,** 1249–1255.

Rebiffe, R., Quillien, J. and Pasquet, P. (1983), *Enquete Anthropometrique sur les Conducteurs Francaises.* Laboratoire de Physiologie et de Biomechanique de de l'Association Peugeo-Renault.

Rebiffe, R., Zayana, O. and Tarriere, C. (1969), Determination des zones optimales pour l'emplacement des commandes manuelles dans l'espace de travail. *Ergonomics*, **12,** 913–924.

Reynolds, H. M. (1978), The inertial properties of the body and its segments. In NASA (1978), op. cit., Vol. I, Chapter IV.

Ridd, J. E. (1985), Spatial restraints and intra-abdominal pressure. *Ergonomics*, **28,** 149–166.

Roberts, D. F. (1973), *Climate and Human Variability*. An Addison-Wesley Module in Anthropology, No. 34 (Reading, MA: Addison-Wesley).

Roberts, D. F. (1975), Population differences in dimensions, their genetic basis and their relevance to practical problems of design. In *Ethnic Variables in Human Factors Engineering*, ed. A. Chapanis (Baltimore: Johns Hopkins University Press), Chapter 2.

Roche, A. F. (1979), Secular trends in stature, weight and maturation. *Monographs of the Society for Research in Child Development*, Serial no. 179; **44**(3–4), 3–27.

Roche, A. F. and Davila, G. H. (1972), Late adolescent growth in stature. *Pediatrics*, **50**, 874–880.

Roebuck, Jr, J. A., Kroemer, K. H. E. and Thomson, W. G. (1975), *Engineering Anthropometry Methods* (New York: Wiley).

Roebuck, Jr, J. A. and Levendahl, B. N. (1961), Aircraft ground emergency exit design considerations. *Human Factors*, **3**, 174–209.

Rona, R. J. (1981), Genetic and environmental factors in the control of growth in childhood. *British Medical Bulletin*, **37**(3), 265–272.

Rona, R. J. and Altman, D. G. (1977), National study of health and growth: standards of, attained height, weight and triceps skinfold in English children 5 to 11 years old. *Annals of Human Biology*, **4**, 501–523.

Rona, R. J., Swan, A. V. and Altman, D. G., (1978), Social factors and health of primary schoolchildren in England and Scotland. *Journal of Epidemiology and Community Health*, **32**, 147–154.

Savinar, J. (1975), The effect of ceiling height on personal space. *Man–Environment Systems*, **5**, 321–324.

Schmidtke, H. (1980), Ergonomic design principles of alphanumeric displays. In Grandjean and Vigliani (1980), op. cit., pp. 265–270.

Searle, J. A., Hardy, R. N. and Haslegrave, C. M. (1980), A geometrical model for the representation of seat-belt fitting problems. *Ergonomics*, **23**, 305–316.

Shapiro, H. (1939), *Migration and Environment* (London: Oxford University Press).

Shotton, M. A. (1984), Problems experienced by rheumatism sufferers in the use of standard seat belts. In *Contemporary Ergonomics 1984*, ed. E. D. Megaw (London: Taylor & Francis), pp. 161–166.

Sigler, P. A. (1943), Relative slipperiness of floor and deck surfaces. National Bureau of Standards Report BM5100 Washington DC. Cited in: Grandjean, E. (1973), op, cit., p. 284.

Simmonds, G. (1979), Ergonomics standards for road vehicles. *Ergonomics*, **22**, 135–144.

Singleton, W. T. (1963), *The Industrial Use of Ergonomics*. Ergonomics for Industry: 1, Department of Scientific and Industrial Research.

Smidt, G. L. (1973), A biomechanical analysis of knee flexion and extension. *Journal of Biomechanics*, **6**, 79–92.

Snook, S. H., Campanelli, R. H. and Hart, H. N. (1978), A study of three preventive approaches to low back injury. *Journal of Occupational Medicine*, **20**, 478–481.

Snyder, R. G., Schneider, L. W., Owings, C. L., Reynolds, H. M., Golomb, D. H. and Schork, M. A. (1977), Anthropometry of infants, children and youths to age 18 for product safety design. Report No. DB-270 277, Consumer Product Safety Committee, US Department of Commerce, Bethesda, MD.

Society of Automotive Engineers (1974), *Devices for Use in Defining and Measuring Vehicle Seating Accommodation*. SAE J8266 (New York: Society of Automotive Engineers).

Sommer, R. (1969), *Personal Space—The Behavioural Basis of Design* (Englewood Cliffs, NJ: Prentice Hall).

Squires, P. C. (1956), The shape of the normal working area. Report No. 275, US Navy Department, New London, CT.

Steidl, R. E. and Bratton, E. C. (1968), *Work in the Home* (New York: Wiley).

Stoudt, H. W., Damon, A. and McFarland, R. (1965), Weight, height and selected body dimensions of adults. *National Centre for Health Statistics, Series 11*, No. 8.

Stoudt, H. W., Damon, A. and McFarland, R. A. (1970), Skinfolds, body girths, biacromial diameter and selected anthropometric indices of adults. *National Centre for Health Statistics, Series 11*, No. 35.

Tanner, J. M. (1962), *Growth at Adolescence* (Oxford: Blackwell).

Tanner, J. M. (1978), *Foetus into Man* (London: Open Books).

Tanner, J. M., Hayashi, T., Preece, M. A. and Cameron, N. (1982), Increase in length of leg relative to trunk in Japanese children and adults from 1957–1977: a comparison with British and with Japanese Americans. *Annals of Human Biology*, **9,** 411–423.

Tanner, J. M. and Whitehouse, R. H. (1976), Clinical longitudinal standards for height, weight, height velocity, weight velocity and stages of puberty. *Archives of Diseases in Childhood*, **51,** 170–179.

Tanner, J. M., Whitehouse, R. H. and Takaishi, M. (1966), Standards from birth to maturity for height, weight, height velocity and weight velocity: British children, 1965, Part 1. *Archives of Diseases of Childhood*, **41,** 454–471; Part 2. *Archives of Diseases of Childhood*, **41,** 613–635.

Taylor, J. H. (1973), Vision. In *Bioastronautics Data Book*, eds. J. T. Parker and V. R. West (Washington, DC: NASA), Chapter 13.

Thompson, D. and Booth, R. T. (1982), The collection and application of anthropometric data for domestic and industrial standards. In *Anthropometry and Biomechanics: Theory and Applications*, eds R. Easterby, K. H. C. Kroemer and D. B. Chaffin (New York: Plenum).

Thomson, A. M. (1959), Maternal stature and reproductive efficiency. *Eugenics Review*, **51,** 157–162.

Tichauer, E. R. (1978), *The Biomechanical Basis of Ergonomics* (New York: Wiley).

Tildesley, M. F. (1950), The relative usefulness of various characters on the living for racial comparison. *Man*, **50,** 14–18.

Travell, J. (1967), Mechanical headache. *Headache*, **7,** 23–29.

Travell, J. E. and Simons, D. E. (1983), *Myofascial Pain and Dysfunction: The Trigger Point Manual* (Baltimore: Williams and Wilkins).

Trotter, M. and Gleser, G. (1951), The effect of ageing upon stature. *American Journal of Physical Anthropology*, **9,** 311–324.

Troup, J. D. G. and Edwards, F. C. (1985), *Manual Handling—A Review Paper* (London: HMSO).

Tutt, D. and Adler, D. (eds) (1979), *New Metric Handbook* (London: Architectural Press).

Van Cott, H. P. and Kinkade, R. G. (eds) (1972), *Human Engineering Guide to Equipment Design* (Washington, DC: US Department of Defense).

Van Wely, P. (1970), Design and disease. *Applied Ergonomics*, **1,** 262–269.

Vihma, T., Norminen, M. and Mutanen, P. (1982), Sewing machine operators work and musculoskeletal complaints. *Ergonomics*, **25,** 295–298.

Vincent, L. M. (1979), *Competing with the Sylph—Dancers in Pursuit of the Ideal Body Form* (Kansas City: Andrews and McMeel).

Ward, J. S. (1971), Ergonomic techniques in the determination of optimum work surface heights. *Applied Ergonomics*, **2,** 171–177.

Ward, J. S. and Beadling, W. M. (1970), Optimum dimensions for domestic staircases. *Architects Journal*, **151,** 513–520.

Ward, J. S. and Kirk, N. S. (1970), The relation between some anthropometric dimensions and preferred working surface heights in the kitchen. *Ergonomics*, **6,** 783–797.

Waris, P. (1979), Occupational cervico-brachial syndromes: a review. *Scandinavian Journal of Work Environment and Health*, **5,** Supplement 3, 3–14.

Weiner, J. S. (1982), The measurement of human workload. *Ergonomics*, **25,** 953–966.

Wells, L. H. (1963), Stature in earlier races of mankind. In *Science in Archaeology*, eds. D. Bothwell and E. Higgs (London: Thames and Hudson), Chapter 39.

Westgaard, R. H. and Aaras, A. (1980), Static muscle load and illness among workers doing electromechanical assembly work. Institute of Work Physiology, Oslo.

Weston, H. C. (1953), Visual fatigue with special reference to lighting. In *Symposium on Fatigue*, eds W. F. Floyd and A. T. Welford (London: H. K. Lewis).

WHO (1980), *International Classification of Impairments, Disabilities and Handicaps* (Geneva: World Health Organization).

Wickstrom, P. (1979), Effect of work on degenerative back disease. *Scandinavian Journal of Work Environment and Health*, **4,** Supplement 41, 1–12.

Williams, J. C. (1977), Passenger-accompanied luggage. *Applied Ergonomics*, **8**, 151–157.

Wilmore, J. H. (1976), *Athletic Training and Physical Fitness—Physiological Principles and Practices of the Conditioning Process* (Boston: Allyn and Bacon).

Wilson, J. R., Cooper, S. E. and Ward, J. S. (1980), *A Manual of Domestic Activity Space Requirements* (Loughborough, Leics: Institute for Consumer Ergonomics).

Winter, D. A. (1979), *Biomechanics of Human Movement* (New York: Wiley).

Wisner, A. and Rebiffe, R. (1963), Methods of improving workspace layout. *International Journal of Production Research*, **2**, 145–167.

Wood, P. H. N. (1975), *Classification of Impairments and Handicaps* (Geneva: World Health Organization).

Woodson, W. E. (1981), *Human Factors Design Handbook* (New York: McGraw-Hill).

Woodson, W. E and Conover, D. W. (1964), *Human Engineering Guide for Equipment Designers* (Berkeley: University of California Press).

Wright, P. and Barnard, P. (1975), Just fill in this form: a review for designers. *Applied Ergonomics*, **6**, 213–220.

Yamada, H. (1970), *Strength of Biological Materials* (Baltimore: Williams and Watkins).

Yamana, N., Okabe, K., Nanako, C., Zenitani, Y. and Saita, T. (1984), The body form of pregnant women in monthly transitions. *Japanese Journal of Ergonomics*, **2**, 171–178 (in Japanese).

Yanagisawa, S. and Kondo, S. (1973), Modernization of physical features of the Japanese with special reference to leg length and head form. *Journal of Human Ergology*, **2**, 97–108.

Index

Bold numbers refer to important sections in the text.